Microphotograph of the erect retinal image in the eye of the firefly (*Lampyris spldl.*). $\times 120$, photographed with Zeiss objective C. The image is of a bay window through which can be seen a church. The letter 'R', cut from black paper was glued to the pane. (It appears as a mirror image due to reversal during printing). The window was 225 cm from the eye; the church is 135 paces from the window.
(The print was made by the Royal Teaching and Research station for Photography and Reproduction in Vienna).

The Physiology of the Compound Eyes of Insects and Crustaceans

A Study by

Sigmund Exner

Foreword by Karl von Frisch

Translated and Annotated by Roger C. Hardie

With 28 Text Figures and 76 Figures in 7 Plates

Springer-Verlag
Berlin Heidelberg New York
London Paris Tokyo

SIGMUND EXNER †
Professor of Physiology
at Vienna University

Translator:
ROGER C. HARDIE
Cambridge University
Department of Zoology
Downing St.
Cambridge CB2 3EJ
England

Title of the German Edition: Die Physiologie der facettierten Augen von Krebsen und Insecten.
Eine Studie von Sigm. Exner (Franz Deuticke 1891)

ISBN-13: 978-3-642-83597-1 e-ISBN-13: 978-3-642-83595-7

DOI: 10.1007/978-3-642-83595-7

Library of Congress Cataloging-in-Publication Data.
Exner, Sigmund, 1846–1926. [Physiologie der facettierten Augen von Krebsen und Insekten. English] The physiology of the compound eyes of insects and crustaceans : a study / by Sigmund Exner ; with a foreword from Karl von Fritsch ; translated and annotated by Roger C. Hardie. p. cm.
Translation of: Physiologie der facettierten Augen von Krebsen und Insekten.
Bibliography: p. 134 Includes index. 1. Insects–Physiology. 2. Crustacea–Physiology.
3. Eye–Physiology. I. Hardie, Roger C. (Roger Clayton) II. Title.
QL495.E9613 1988 595.7'01'823–dc1988–29450

Softcover reprint of the hardcover 1st edition 1989

Data conversion: Daten- und Lichtsatz-Service, Würzburg

2131/3130-543210 – Printed on acid-free paper

Dedicated to
HERMANN VON HELMHOLTZ
on the occasion of his 70th birthday,
31 August 1891

Foreword to the English Edition

It is remarkable that a fundamental scientific text from the year 1891 can be reprinted today without any of its basic points having been proved wrong in the intervening years. Because of its international significance the new edition is no longer in German but has been translated into English. The author, Sigmund Exner (1846–1926) was Professor of Human Physiology at the University of Vienna. Mainly due to the present monograph he counts as one of the founders of comparative physiology.

Exner's study of compound eyes was stimulated by the relative ignorance of their physical properties and by the enormous variety in structure and function compared to the more uniform lens eyes of the vertebrates. For example, in truly diurnal insects the retinal elements directly adjoin the dioptric apparatus, each ommatidium is cloaked in a pigment sheath and is optically isolated from its neighbours.

Like the stones of a mosaic, the image points provided by each ommatidium build up an upright image. In nocturnal animals the retina lies a certain distance behind the crystalline cones which now have a different construction leading to a new ray path: making use of the greater distance from the retina, each retinal element now receives light from a whole group of cones so that the intensity of the image is increased. Exner called this mechanism the superposition image in contrast to the apposition image of diurnal animals.

However, the superposition of the images is not perfect, and leads to a blurred image. With increasing intensity the eye corrects for this defect by means of a redeployment of the pigments in the retina. These distribute themselves so that the light leaving the crystalline cones is prevented from crossing over into neighbouring ommatidia, thus converting the superposition eye into an apposition eye. This renders the image dimmer (which is permissible under the higher intensity) but brings about an increase inacuity – a truly sophisticated biological control mechanism!

Exner himself did not continue his studies on pigment migration. However, since he was my uncle, I knew him well, and thus it transpired that in 1908, as a 22-year-old student, I started my first major scientific project in his exemplary institute. As a project he suggested I should investigate the mechanism of pigment migration in compound eyes, and in particular which was the resting and which the excited state of the pigment cells. Unfortunately, my attempts to determine this were inconclusive. Exner himself was disappointed by this negative result, yet for me the search for a practicable solution under Exner's supervision was a rewarding and exciting experience. The question of the mechanism of pigment migration in compound eyes, like many other mysteries of this marvellous visual organ, remains unsolved to the present day. Sigmund Exner was so fascinated by these eyes that it was with regret that after several years he forced himself to leave the subject and return to his real field, human physiology. We are eternally grateful for this by-product of his research which time has proved immortal.

Special thanks are due to the Springer-Verlag for their willingness and co-operation in producing this new edition.

14 March 1982 Karl Ritter von Frisch

Preface to the German Edition

The present study, which is the result of several years work, captivated me more than one might have expected of a physiologist – the compound eye lying somewhat off the beaten track from our normal research. However, there is an undeniable fascination to the question of how and why Nature has found two so fundamentally different methods to arrive at apparently the same result – namely to equip creatures with organs of vision. Further, once I became more familiar with the subject, I found such a wealth of forms and phenomena which posed question after question demanding such a variety of anatomical, biological, physical and physiological approaches, that my interest was always reawakened. The invertebrate eye is a Proteus compared to that of the vertebrates, which might indeed appear boring to someone acquainted with the former.

Thus it was with some regret that I had to tear myself away from my object of study, leaving a number of questions unanswered. How tempting it would have been to include the eyes of the myriapods and the arachnoids in the study, or to study the peculiar eyes of the crustaceans from the deep sea fauna. Thus in the following pages, certain points are only mentioned in passing or left unanswered. For example, the experiments on the mechanism of pigment migration have not progressed beyond the stage of pilot experiments.

To orient myself in a novel field, it was often necessary to seek the help of colleagues. I acknowledge the most cordial support of the staff of the Vienna Natural History Museum, particularly Professors F. Brauer, Ganglbauer, Kölbl and Rogenhofer and Director A. Brezina. I thank Professor Claus and Grobben for sending me live marine specimens from Triest, and Professor Ed. Suess for specimens of fossilized crustaceans. I am particularly indebted to the management of the zoological station in Naples, which, apart from scientific support, provided not only the most abundant material, but also the necessary equipment and facilities for the physiological experiments during my stay there. The wealth of forms which I came to see there would probably provide work for decades to come. My warmest thanks, therefore, to Professor Dohrn and the other staff of the institute!

Finally, I wish to thank the publishers, F. Deuticke, who, although they cannot have reckoned with a large readership, have produced, in every respect, an immaculate copy, respecting all of my wishes regarding the presentation of line drawings, colour plates and photographs.

Vienna, 22 March 1891 — Sigmund Exner

Preface

There can be few other fields of research where a single study has been of such lasting significance as is the case with the monograph of Sigmund Exner. This book was quite literally decades ahead of its time and has been the cornerstone of physiological optics in compound eyes for almost a century. Even today, hardly a serious paper on the subject appears without it being referenced. Despite its undoubted importance, however, the monograph has also almost certainly been largely appreciated only at second-hand. Apart from the scarcity of copies, the fact that the original version appeared in the German language has made it effectively unavailable for a large section of the present-day scientific community.

In 1980, when I was employed at the Max-Planck Institute for biological Cybernetics in Tübingen, I was approached by Doekele Stavenga with the idea of translating Exner's text, for the first time, into English. Although the task has turned out to be a lengthy one, it has been far from dull. Like most other English-speaking compound eye researchers, I had also never read the original text, and was constantly surprised by the depth of insight shown by Exner with what today must appear the most primitive of techniques. His major contributions – the resurrection of Müller's mosaic image theory, the discovery of the lens cylinder principle and the mechanism of superposition image formation – were of course familiar, but buried within the book are numerous other gems which I hope give the reader as much pleasure as they did me.

By and large I have translated the text faithfully, trying to retain the original flavour, though on occasion I have shortened some redundant passages. I have annotated the text only sparingly, in particular giving answers to questions raised directly or indirectly by Exner himself and pointing out the few places where he actually made mistakes. I have not attempted to provide a comprehensive bibliography, but a list of the most important recent review volumes is included for further reference. In particular, the book "Facets of Vision" * is the most recent attempt to review the entire field of compound eye research, and is the proceedings of a conference held in 1987 in Regensburg, West Germany to commemorate a century of compound eye research since Exner.

* Stavenga DG, Hardie RC (eds) (1989) Facets of Vision. Springer, Berlin Heidelberg New York Tokyo

Thanks are due to Professor Klaus Vogt for help with some of the trickier passages and Dr. Daniel Osorio for reading the manuscript. Dr. Dan-Eric Nilsson very kindly gave me permission to reproduce his figures illustrating the different optical mechanisms utilised by compound eyes (Appendix I) and his expert knowledge was invaluable for identifying the eye types of all the species used by Exner (Appendix II). I also most particularly wish to thank Dr. Doekele Stavenga, who provided the impetus and conscientiously checked my first draft of the translation, and finally Professor Hansjochem Autrum and the Springer-Verlag, whose interest in the project allowed it to get off the ground.

Cambridge, September 1988 Roger Clayton Hardie

Contents

The Author, Sigmund Exner (1846–1926)

Chapter I Physical Principles

For my first studies on insect eyes (Exner 1875) I attempted to determine the refractive index of the cornea in *Hydrophilus*. This consists of numerous cylinders whose length is three to four times their width. The front surface, which in vivo may be in contact with either air or water, is weakly curved whilst the inner surface has a very small radius of curvature (Fig. 23). In such a case one normally determines the refractive index by measuring the radius of curvature of both surfaces, the distance between them and the distance of the image from one of the surfaces. The resulting refractive index turned out to be exceptionally high – namely 1.8.

With the help of a microrefractometer (Exner 1885) I later realized that the corneal cylinder was not, as assumed in the previous calculations, homogeneous, but rather was composed of cylindrical layers with refractive index decreasing gradually from the axis towards the margin. I further found that a fragment of such a cylinder had a refractive index of only 1.55 (similar to normal glass) and that the corneal cylinder could still produce inverted images of external objects after I had cut away both surfaces.

The erroneous refractive index originally determined thus came from the fact that I attributed the measured image distances solely to the focusing power of the two surfaces without considering the, at that time, unsuspected focusing power of the layered cylinder.

Layered cylinders, with refractive index decreasing continuously from the axis towards the periphery, function in certain respects like lenses; I therefore call them lens cylinders. In some respects their properties differ considerably from those of a lens. As will be shown later, they play an important role in compound eyes, having optical effects which could not be achieved by normal lenses. Although a detailed account has been published previously (Exner 1886 a, b) a brief description of their properties is repeated here (see further K. Exner 1886, Matthiesen 1886).

Let there be a cylinder *abcd* (Text Fig. 1) with refractive index having a maximum along the axis *xy* and decreasing steadily towards the margin. The two ends *ac, bd* are flat surfaces perpendicular to the axis. *xm* is a light ray; as soon as it enters the cylinder it encounters boundaries between layers of decreasing refractive index, *n*. At each such boundary, e.g. *a'b'*, it will be refracted away from the perpendicular (*pq*) so that its path makes a steadily decreasing angle with the axis. Eventually this angle reaches zero and then becomes negative. Since the ray now passes into layers of higher refractive index, it will now be refracted towards the perpendicular (*p'q'*) and thus intersects the axis again at *y*. From symmetry all rays leaving *x* with the same angle will meet again at *y*.

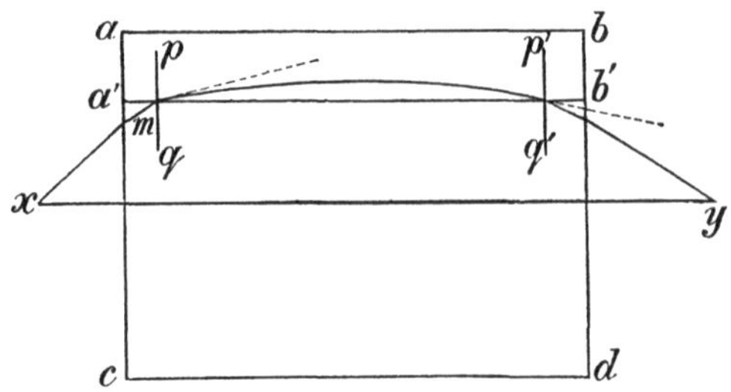

Text Fig. 1

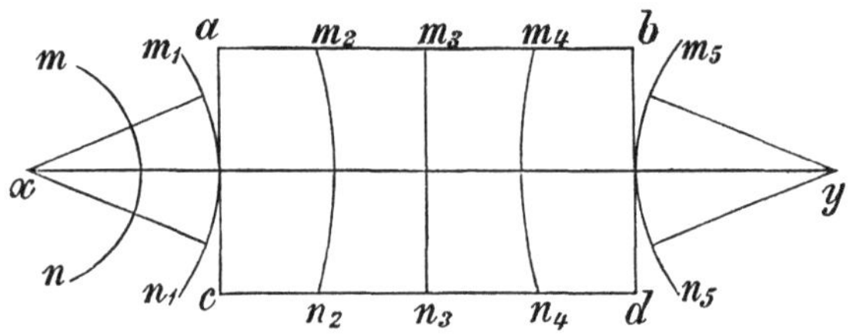

Text Fig. 2

At first sight it might appear that, once parallel to the axis, the ray should remain so, i.e. all rays should emerge parallel to the axis. A more exact consideration shows that this is false. One only has to consider the incident ray bundle in terms of its elementary waves and it is then clear that those located nearer the axis must have a lower speed of propagation. We find ourselves here at the limit of geometrical ray optics.

To see if rays from x striking the cylinder with a different angle of incidence also converge on y requires calculation which was first performed by my brother Prof. Karl Exner. The conclusion is that when one only considers the paraxial rays (as is also the case for calculations on normal lenses) then these rays do in fact meet at y. If other rays are also to meet at y, then the refractive index gradient must follow a parabolic function.[1]

One can also imagine the process as follows: again *abcd* is the cylinder (Text Fig. 2), *mn* is the surface of a spherical wavefront emerging from a point source, x. As this reaches m_1n_1 it starts to undergo distortion such that the speed of propagation is slowest along the axis. It changes gradually through m_2n_2 m_3n_3 etc. and emerges once more in the air as a concave wavefront m_5n_5, i.e. the rays emerging from the cylinder are convergent. From Text Fig. 2 one can also see directly that if the cylinder is cut short at m_3n_3, the emerging wavefront is flat and perpendicular to the axis, i.e. that then x would be the primary focal point of the cylinder. The construction of the secondary focal point follows similarly: when a plane wavefront (i.e. one originating from an infinitely distant point) falls on the cylinder m_3n_3bd, then the wave emerges transformed into a spherical wavefront m_5n_5, i.e. y is the secondary focal point of this cylinder.

The wave optics treatment also offers the simplest proof that, when only paraxial rays are considered (as with spherical lenses) then an image of x must be projected irrespective of the function according to which n decreases from the axis. This function determines rather the shape of the wavefront m_5n_5. In any case

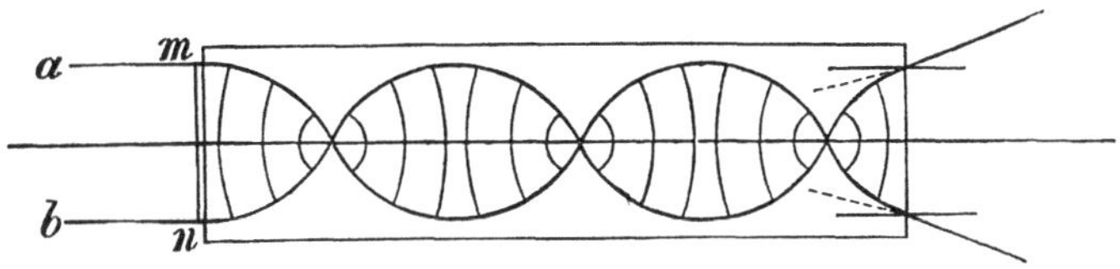

Text Fig. 3

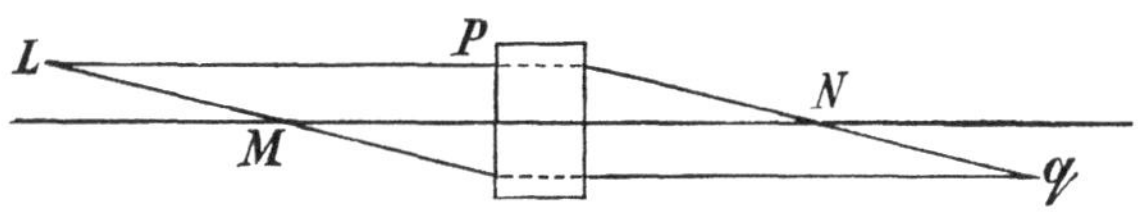

Text Fig. 4

though, because of the rotational symmetry of the cylinder a small part of this wavefront will also have rotational symmetry, i.e. it can be approximated by a spherical surface. The centre, *y*, of this spherical surface is thus the image of *x*.

Calculation shows that the focal length of the cylinder is

$$p = c/l,$$

where *c* is a constant and *l* is the length of the cylinder. The optical power can thus be tailored according to the cylinder's length.

However, the formula is only valid within certain limits. In fact the calculations are only simple to perform when the length of a cylinder – with a given refractive index gradient – is sufficiently small. Apart from the calculated case, intuition shows that the focal length is a periodic function of cylinder length. For example the behaviour of a parallel ray bundle must approximate that shown in Text Fig. 3, where *mn* represents the incident plane wavefront and *ab* two rays thereof. The deformation of *mn* is indicated. In such a cylinder there is thus a succession of foci and as its length increases it functions alternately as a convergent then a divergent lens.

With comparatively short or weak lens cylinders, and if the object to be focused is not a point, then images can be constructed in the same way as with lenses. Just as the rays coming from one point on the axis converge on one point, the rays from a point neighbouring the axis will also converge approximately on one point. In order to find this point of convergence only two rays are needed – the focal ray and the parallel ray. If *M* and *N* (Text Fig. 4) are the two focal points and *L* a point near the axis then the ray *LP* will be refracted through *N* and the ray *LM* through *Q* so that the image of *L* lies in *Q*.

With respect to insect eyes, we are mainly interested in lens cylinders of two lengths, first those whose focal plane lies approximately at the back surface, and secondly those where the focal plane lies in the middle of the cylinder.

A. Focal Plane in the Back Surface of the Cylinder

Rays incident parallel to the axis intersect where the axis meets the back surface. Thus in Text Fig. 5 *abcd* is a cylinder with axis *xy*; an incident ray bundle *mn*,

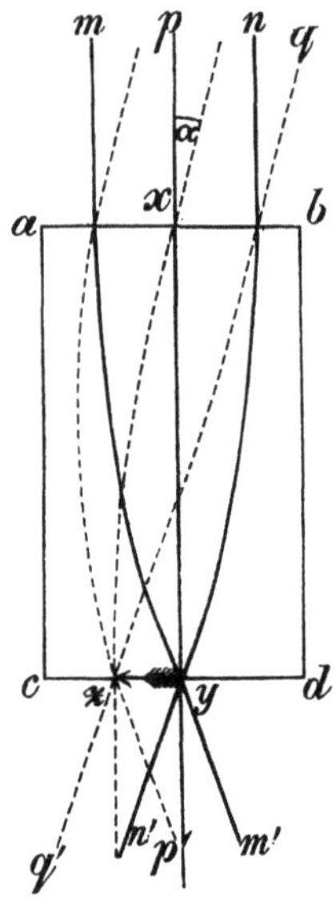

Text Fig. 5

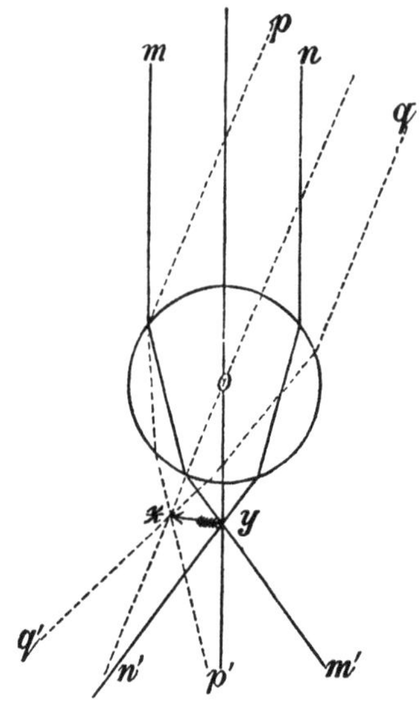

Text Fig. 6

parallel to the axis converges at the focal point *y*. The image of another point, the ray bundle from which is indicated by *pq*, lies in *z*. An inverted image of the external image thus exists in the back surface of the lens cylinder. In certain points, this image differs considerably from that produced by a lens. For example in Text Fig. 6 *yz* is an equally large image of the same object produced by a spherical converging lens. One sees that the principal rays passing through *o* maintain the same angle with each other after refraction; in other words the cones of light formed by rays from two different points have different axial directions. In the lens cylinder, however, the cones are parallel. It is thus clear that in order to utilize the light from a particular direction, e.g. *xy* and the direct surroundings up to *a* (Text Fig. 5), most profitably (i.e. to direct it towards nerve endings), the lens cylinder is superior to the lens.[2]

If we call any ray which hits the middle of the front surface of the lens cylinder a principal ray (by analogy with normal lenses) the difference between the two optical arrangements can be expressed as follows: in the case of the lens the principal rays from different points on the object diverge after refraction, but in a lens cylinder (with focal length equal to its own length) all the principal rays travel parallel to the axis. One sees directly from Text Fig. 5 that the size of the object that can be imaged in *cd* is strictly limited by the thickness (*cd*) and length (*ac*) of the cylinder (whilst with a spherical lens the size is unlimited), and that for the optical function the proximal outer layers of the cylinder are irrelevant. These could thus be absent, i.e. a truncated cone (the basic shape of cone elements in compound eyes) would have the same effect. A truncated cone with length equal to its focal length and built on the principle of a lens cylinder thus focuses all the light from a restricted (axial) field of the visual world onto its back surface in such a way that all the principal rays leave the cone parallel to its axis.

B. Lens Cylinder Twice as Long as its Focal Length

In this case, of course, the inverted image of an object at infinity lies in the middle of the lens cylinder (*yz* in Text Fig. 7). If the lens cylinder *abcd* were cut at the level *yz* then we would again have the previous case (A) and all the principal rays would emerge parallel to the axis. A second lens cylinder with identical optical properties now starts at the level *yz* and extends to *cd*. *yz* serves as the object for this second half cylinder. According to the general principle of reversibility, the ray path in the second half of the cylinder must be symmetrical to that in the first. Thus rays (*mn*) which enter the cylinder parallel to the axis meet at one point *y*, and will then behave in exactly the same way in the lower half of the cylinder, only their direction of travel is opposite about point *y*. They must leave the cylinder parallel again, just as they entered it (*m'n'*). A ray from a neighbouring point (*pq*) striking the cylinder with an angle α to the axis, must also leave the cylinder with the same angle ($\beta = \alpha$). The statement above, that the principal rays are perpendicular to the plane *yz*, is a logical consequence.

A lens cylinder with such optical properties constitutes a non-magnifying astronomical telescope focused at infinity. In principle one can achieve the same effect with a combination of two similar convex lenses separated by twice their focal length. Here as well, the emerging ray makes the same angle with the axis as it made before entering the lens combination; axis, incident and emergent rays all lie in the same plane, and the two rays on the same side of the axis.

The examples of lens cylinder properties discussed above are amongst the simplest imaginable and are probably never realized in compound eyes. Here, as

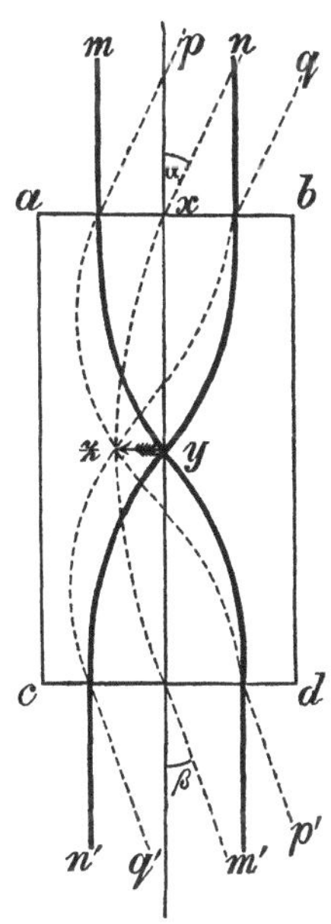

Text Fig. 7

a rule, one finds instead refractive effects achieved through the combination of spherical surfaces and lens cylinders. Any spherical boundary between media with different refractive indices projects an image of an external object; the lens cylinder does the same. If both happen to function as converging lenses and the spherical surface at the base of the lens cylinder is suitably situated, then the two kinds of refraction complement each other, just as the focusing power of one convex lens is complemented by a second. This is naturally also the case when the lens cylinder is bounded by spherical surfaces on both sides, and is similar to the situation (as in most arthropods) where the dioptric apparatus consists of two parts, the cornea and crystalline cone, each of which probably represents a lens cylinder with both ends replaced by spherically curving surfaces.

Compound eyes also show other differences from the the simple physical principles described here. In Text Fig. 7 it is assumed that the two halves of the cylinder (above and below *yz*) have exactly the same optical structure, resulting in the effect of a non-magnifying astronomical telescope. In Nature it appears to be the rule that the two parts are of unequal optical structure, as in the case of a magnifying astronomical telescope. The image *yz* then no longer lies in the centre of the cylinder but nearer to one end. The rays from one point can nevertheless still leave the optical system parallel.

It is possible that yet another kind of image formation plays a role in insect eyes. Recently Matthiesen (18–, 1886) has pointed out that an array of spherical shells, all with parallel boundaries, also has the effect of a converging lens when the refractive index of the shells decreases in the direction of the light path, and when the surfaces are concave with respect to the incident light. Such a lens is still effective even when both ends are plane surfaces, thus forming a cylinder.

Matthiesen cites a number of insect eyes that he supposes to work by this principle, which he calls the "step lens", his reasons being that Grenacher's studies have shown the cornea to be appropriately layered, and the more proximal layers are assumed to to be less optically dense as they are younger. Due to the frequent occurrence of such layering it is not improbable that such a principle operates in insect eyes – as already admitted – however, only relatively small effects can be achieved in this way. The strong refraction shown by insect eyes would require such large differences in refractive index between neighbouring layers that they would have to be directly detectable in the microscope, but this is not the case. The situation is different for a combined step lens and lens cylinder, which Matthiesen (1886) also discusses, and whose principle should be made clear in Text Fig. 8. Here there are hyperbolic layers with refractive index decreasing in the direction of the incident light (*ab*). One sees at once that the structure of a lens cylinder is thereby automatically generated – since in any cross-section the refractive index decrease from the centre to the periphery. It is exactly this kind of layering which stands out, for example, in the *Limulus* eye; I therefore consider it very probable that it also plays a role elsewhere. As we shall see later (Chap. II) the properties of this layering are – as to be expected – basically those of a lens cylinder.

I referred above to the differences in optical properties of spherical boundaries and lens cylinders. Looking at the various arthropod families, it appears that evolution favours an ordinary lens in some cases and a lens cylinder in others.

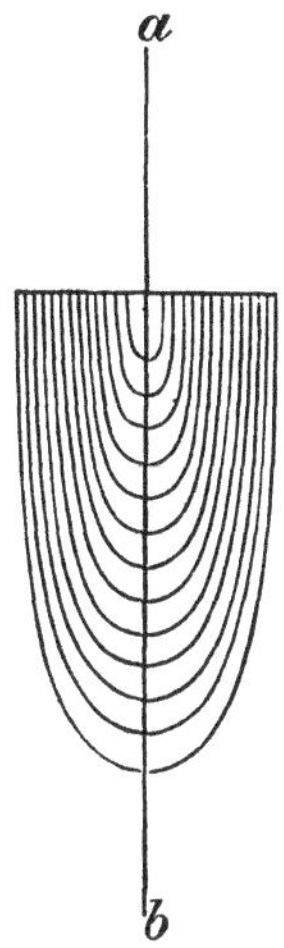

Text Fig. 8

This accounts for the great diversity amongst compound eyes, in contrast to which the simple eye with an inverted image shows a pitiful uniformity.

I know of no compound eye where the principle of the lens cylinder alone dominates, and equally of no compound eye where the lens principle alone controls the optics. Indeed, even in the vertebrate and human eye the lens cylinder is utilized (cf. Brücke 1887) in that the lenses have a layered structure and in their centre a cylinder-like structure whose axis more or less coincides with the eye's axis.

The action of a lens cylinder is virtually independent of the surrounding medium which, however, critically affects the action of a spherically curved surface. Correlated with this is the fact that amphibious animals, e.g. water beetles, have a very weakly curved front surface, whereas, for example in many butterflies, this surface has a very small radius of curvature. On leaving the water the optical effect of a corneal facet would completely change if it were strongly curved, whilst it would hardly change at all if it consisted of a lens cylinder. In the case of crustaceans which temporarily leave the water a similar situation prevails.

Another factor determining the diversity of eye forms is that the type of lens cylinder discussed above need not be the most advantageous for different life styles. Thus, for simplicity, I have assumed that the lens cylinder has exactly the same optical structure for the whole of its length. It could, however – and this in fact occurs – have a steeper refractive index gradient in the proximal portion than distally, there being a gradual transition between the two parts. It is then equivalent to a magnifying astronomical telescope focused at infinity, i.e. the inverted image no longer lies in the centre but is displaced towards the back surface. But it still lies in the secondary focal plane of the first lens cylinder and in the primary focal plane of the second, when we again imagine the cylinder cut into two parts at the level of the inverted image.

Chapter II Dioptrics of the Compound Eye

Historical Background

In 1826 Müller presented a theory of insect vision according to which the retinal image should be upright. In contrast to the vertebrate eye, he supposed the retinal image to be achieved, not so much by the convergence of rays from each point on the object, but rather by the separation of rays from different points on the object.

Müller had in fact realized that the so-called compound eyes consist of a large number of elements (ommatidia), each representing a black pigmented tube. These tubes are arranged radially on the surface of a more or less perfect hemisphere. Whatever the contents of the tube, as long as they are transparent, an upright, though not necessarily perfect, image of an external object must be projected onto the surface of the hemisphere, since it is obvious that only rays which are incident approximately along the axis of each tube can reach its base. Rays entering with a greater inclination to the tube's axis will strike the sides of the tube before reaching the base and will be absorbed by pigment. If there is a nerve end organ at the base of each tube, i.e. if the hemispherical surface is composed of a light sensitive nervous array, then this convex retina will function in the same way as the concave retina of the vertebrate eye.

This is the essence of Müller's theory of mosaic vision and the upright retinal image. Later, it was subsequently abandoned and in fact almost forgotten following the work of Gruel (1844) and Gottsche (1852). They made the readily repeatable observation that, under suitable conditions, an inverted image of an external object could be resolved with a microscope under each facet of a fly's eye. Such images were in fact already known to Leeuwenhoek and others. A remark made by Müller in response to Gottsche's paper may have given the reader the impression that he was prepared to drop his theory of the upright retinal image in view of the visible inverted "micro-images". There followed a number of comparative anatomical and physiological studies which ignored Müller's theory on the evidence of these inverted images. This is quite remarkable, since the investigators involved were almost exclusively microscopists who must have been aware that every oil-drop, air bubble etc. projects a microscopic image; it should thus not be expected that the demonstration of such an image behind each facet should be attributed such great importance in refuting Müller's theory – all the more so when one considers the questionable conditions under which Gottsche demonstrated his inverted images[3] (for further discussion, see Exner 1875).

And so, in the year 1868, Schultze was able to say (with reference to the experiments of Gottsche and Zenker): "the physically untenable theory of the mosaic, upright image in the insect eye is thus experimentally disproved" – and he then undertook the thankless task of trying to locate the retina corresponding to the hypothesized inverted retinal image in each ommatidium. The first turning point came 19 years after Gottsche's publication. Boll, a student of Schultze, inspired by the observation that the rods in the retina of triton projected inverted images, questioned the functional significance of the facet images and urged for a return to Müller's theory (Boll 1871).

Later, independently and using different approaches, first Grenacher (1877, 1879), and then myself (Exner 1875) both further supported Müller's theory of the upright image. On the basis of his extensive and successful investigations on simple and compound eyes of numerous lower animals, and especially from his thorough knowledge of the relevant nervous elements of the retina, Grenacher became convinced that the theory of individual images was untenable, that even if there were such images there could be no retina that was suitable for their physiological evaluation, and that the anatomical situation was altogether in favour of Müller's theory. In a certain sense I followed the opposite path. Thus I started with the supposition that the basic optical principles in similarly constructed eyes would also be similar, and thus investigated the eye of one animal (*Hydrophilus piceus*) in detail. I was able to show that Gottsche's images were indeed visible when one proceeded as he had done, but that such an image was not produced in the living animal and further that it is absolutely impossible for an image to lie where, according to his theory, it would have to in order to be perceived. I believed instead to have shown that the function of the rather complicated dioptrics was to guide rays, which are incident approximately parallel to the ommatidial axis, to the tip of the crystalline cone, partly by refraction and partly by reflection. Here they can excite the nervous elements much more intensely than if this dioptric/catoptric apparatus were absent. In this way the intensity of the upright retinal image is increased as Müller had already supposed, although his reasoning was different, as seen in the following, physically somewhat confused passage (Müller 1826):

"The convexity of the corneal facets will guide axially incident light towards the axis by refraction, resulting in greater convergence in the depth of the eye. The light passing through the cone might thus be focused to a point at its tip where the visual fibre makes contact. This must result in the resolution of the image being greatly increased. However, the refraction due to the external convex surface of the cornea is not so great that separate small images could be produced under each facet."

In the eye of *Lampyris splendidula* I was able to demonstrate the concentration of rays at the tip of the crystalline cone with complete confidence. This does not arise, as Müller proposes, simply through refraction at the corneal surface (in this case at least an indistinct inverted image would have to result) but – as I then believed – through total reflection at the margin of the crystalline cone. Thus, as anyone can imitate with a drawn out glass rod, once light is caught in the cone, it is further guided to its tip. I later found an optical mechanism in the firefly eye, namely the layered lens cylinder, which as far as the concentration of rays at the tip of the cone is concerned, produces the same effect as internal reflection but which is based upon refraction. The reason for using the firefly eye for these experiments is that the crystalline cones and the cornea are fused. One is thus in the fortunate situation of being able to brush away the pigment and other soft parts of the eye and to investigate the entire dioptric apparatus with the normal juxtaposition of the cones and facets.

I also pointed out that the results of my optical investigations of the insect eye could provide the key to the explanation for the empirical observation that insects recognize their friends and foes much more by their movement than from their shape.

In certain animals, Schmidt (1878) found asymmetrical crystalline cones showing diverse irregularities, in particular horn-shaped bends. Remarkably, from this he concludes that not only is the theory of inverted images untenable (with which everyone would agree, assuming that his observations are correct), but also the theory of mosaic vision. Grenacher has already shown him wrong in the latter respect so that I can limit myself to the following remarks. Schmidt has himself prepared bent glass rods and cones similar to his crystalline cones to see how light is propagated in them.* He also supposes that light rays behave similarly in the eyes he discusses. But he does not seem to have realized that, even with such a system, mosaic vision is still possible. Provided that both the light-receiving parts of the crystalline cones and their tips are arranged in a radial mosaic, then under appropriate conditions an image according to Müller's principle must result whether the cones are bent or not and whether they are all bent in the same or different ways.

With respect to the upright retinal image, Notthaft (1880) also favoured Müller's theory and put forward the following view on the functioning of compound eyes:

"A cylindrical light beam, or line of light, is incident on every retinula[4] precisely along its optical axis. The individual rays comprising this beam maintain exactly the same direction. The part of the object from which these rays originate, and which fills the visual field of one element is thus always the same size whatever the distance; namely, it is exactly equal to the cross-section of the proximal, non-pigmented tapered end of the cone, or the cross-section of the retinula."

He elaborates further, that since the rays of each beam are strictly parallel, they must always originate from the same number of neighbouring point sources in the outside world, collectively forming an elementary visual field, whether the perceived object is close or far away. Just as the intensity of a point source decreases with the square of the distance, so must the intensity of the entire elementary visual field. The insect thus orients itself by seeing all distant points dark and close ones bright. Having no eyelids, it would even be able to expose its eyes to direct sunlight since the sun is so far away, etc.

Apart from the fact that the insect would perceive every dark object as distant, this theory is also untenable on physical grounds. Thus if, by what is referred to as a light beam or line, is really meant an infinitely thin ray, then this must originate from an infinitely small area of the object, i.e. a point. Then unless the

* Since my first investigations on the compound eye I have demonstrated a number of such glass rods and cones, some with complicated bends, to show how light is propagated within them. Recently this very old experiment has found practical application – when it is necessary to guide light "around the corner", and when lens and mirror cannot be used. Although such a "light guide lamp" has been brought onto the market, for the illumination of microscopic objects lens and mirror will probably always be preferable.

brightness of the object is infinite there will be an infinitely small amount of light (i.e. none) originating from each point of the object. In other words, the insect could not see. On the other hand, if, by the light beam, a number of parallel rays with finite cross-section is understood (as is probably assumed in the above), then these rays can never remain parallel when they pass through curved boundaries between different refractive media such as those found in the ommatidia of insect eyes, and which Notthaft of course also had recognized. However, if the rays are refracted the laws of refraction come into force, and the brightness no longer decreases with the square of the distance. Notthaft has made exhaustive measurements of the size of the corneal facets in numerous insects, which in themselves are valuable. Whether he is right in considering the size of a facet as a measure of the acuity of the animal remains to be seen – I am unable to see any justification for this.[5]

A peculiar modification of Müller's theory originates from Thompson Lowne (1884). He supposes that the structure known as the retinula (which is generally considered part of the nervous part of the visual organ) is in fact part of the dioptric apparatus, and the retina (as yet overlooked) lies behind the retinulae. He goes on to say that the retinula actually has a different shape and properties than usually imagined – it should represent a distally convex, rather voluminous object, which, by virtue of its convex surface, can project an image of an object lying in front of it onto the retina. Gottsche's inverted images in the crystalline cones, seen in every facet with the microscope, then serve as the object. The result is a twice-inverted, i.e. upright, image of an external object on the presumptive retina. Each such image comprises only a small part of the visual field so that an upright retinal image is constructed, the individual parts of which, as numerous as the facets of the eye, are themselves upright images of the respective portions of the visual field.

I cannot agree with this hypothesis either. Thus firstly I can see no optical medium in the retinulae capable of projecting an image; alone the presence of rhabdoms with their complicated cross-sectional shape seems to me to make this impossible. Secondly I cannot agree, as Thompson Lowne assumes, that the retina lies behind the retinulae. Of course it is very dangerous to make precise statements about the location of the retina in compound eyes, just as the question of the position of the sensory nervous structures in the vertebrate eye is not yet positively answered. However, from the behaviour of the retinal pigments, particularly the tapetum, in compound eyes (see Chap. V) I believe that one can conclude with near certainty that light produces nervous excitation in one of the layers distal to the proximal ends of the retinulae. Thus no one would assume that, at the onset of darkness, a thick layer of strongly reflecting tapetal substance should move between the dioptric apparatus and the light-sensitive layer in nocturnal animals. It would be much more reasonable when, with respect to the light path, the tapetum lay behind the sensitive layer, as in mammals. I refer again later to the pictures which Thompson Lowne used for the postulation of his retina.

By and large it seems that Schultze was the last to look down upon Müller's theory with a superior smile, and in recent years investigators unquestionably favour it. I cite, for example, the beautiful observations of Forel (1886), which are firmly based on this theory, also those of Plateau (1887, 1888), and the arguments of Sharp (1889). From his investigations of the *Phronima* eye Claus (1879) also comes down against the Leydig-Gottsche theory and in favour of Müller. Recently Patten (1887) has published anatomical findings which are not in support of Müller's theory, but these are much in need of corroboration.

Finally, for the sake of completion, I should mention that at the 61st meeting of German biologists and doctors in Cologne, I demonstrated the upright retinal image in the firefly eye under the microscope (Exner 1889a).

My findings suggest that compound eyes can be divided into three types according to their optical mechanisms. Two of these function dioptrically, the third predominantly catoptrically. I call the retinal images of the first two types the apposition image and the superposition image according to their means of formation. (The latter is identical to the image in the *Lampyris* eye which I previously called the summation image. Now that I have become aware of the apposition image I find it better to abandon this name so as to avoid misunderstanding).

In order to study the dioptric apparatus of the eye, it is highly desirable to work with the intact system. Most eyes do not meet this requirement, commonly the crystalline cones are attached to the cornea by soft tissue and easily become detached as soon as one dissects the eye. The pigment further renders the exact study of the optical properties impossible.

However, there are animals where the crystalline cones, or analogous cuticular structures with similar function, are fused to the cornea. Here the entire dioptric apparatus is in one piece, it can be freed from the pigment and its optical properties studied. For the study of the two dioptric eye types I chose: the horseshoe crab (*Limulus*), which also has the advantage of having very large ommatidia, and the firefly, which is admittedly small but easy to procure in sufficient numbers.

A. The Apposition Image of Limulus

The large, completely immobile eye of the horseshoe crab has a very large radius of curvature. The overall shape is lenticular, the long axis being horizontal. The surface is robust and chitinous and the facets can even be recognized with the naked eye. Even in dried specimens, though better in specimens preserved in alcohol and best in the living animal, one can see a dark spot lying in the depth of the eye which gives the perfect impression of a large pupil. This "pseudopupil" is only an optical phenomenon, as one sees at once when one turns the whole animal around. The pupil then maintains its position such that it is always directed towards the observer. This pseudopupil can be seen in many species and I will return to it later. Leydig (1855) has also described it and recognized it as an optical phenomenon.

In a section perpendicular to the surface one sees numerous chitinous cones adhering to the cornea and projecting into the depth of the eye (Fig. 20). One can hardly detect any curvature in the corneal facets, further the apical surfaces of the truncated cones, at least in some parts of the eye, even appear slightly concave.

As Grenacher (1879) illustrates, the entire structure is laminated from the outer to the inner surfaces. Moreover, certain layers stand out against their surround ings by virtue of their optical properties: a corneal layer already showing clear protrusions, and a layer separating the cone margin from the cone interior. The latter, which is not parallel to the lamellae, is not always so distinct but is probably present in every cone (Text Fig. 10). It gives the vivid impression of a cone of different optical properties lying within each cone. The cone margin is not anatomically distinct from the core, only differing in terms of its refractive index. At the interstices between the cones, the chitinous substance is penetrated by very fine pore canals perpendicular to the surface (Fig. 20, *a*) best seen in horizontal sections. (In the figure only a few, air-filled canals are shown; most are fluid-filled and hence indistinct).

In many sections I could see neither the inner corneal boundary nor the separation of the cone into two components and I thus suppose that the underlying optical differences are differently pronounced in different parts of the eye.

A conspicuous characteristic of the *Limulus* eye, which severely hinders the optical investigation, is that only the cones in the central eye region have axes perpendicular to the corneal surface. The cone axes do not converge upon the centre of curvature of the cornea, but rather on points lying nearer the cornea. Thus as one approaches the periphery of the eye the cones are increasingly skewed with respect to the cornea. The angle between the cone's axis and the cornea can deviate from the perpendicular by at least 40° (Text Fig. 9).

Apart from its truncated apical surface the entire cone is enclosed in black pigment. In my sections the retinulae (*R*), with their star-shaped rhabdoms, lie at a distance of ca. 0.04 mm from the cone tips. In comparison with other eyes the retinula elements are exceptionally short compared to the dioptrics. The retinula is also enclosed in pigment which is practically continuous with that of the cone so that only a short zone between the cone tip and the nervous elements remains free of pigment.

The apical surface of the cone (by this I mean the truncated surface) is usually elliptically, rather than circularly, rounded. In one example I found the long diameter to be 0.09 mm and the short diameter 0.07 mm. There may be some variation in this, e.g. Grenacher quotes a diameter of 0.065 mm.

It is naturally desirable to have fresh animals at one's disposal and thus I had two live specimens of *L. polyphemus* delivered to Vienna from the Berlin Aquarium. If one dissects out the eye it is easy to peel away the dioptric apparatus from the soft parts, and careful brushing removes any remaining pigment.

To gain a proper idea of the functioning of such a dioptric apparatus, its surfaces should be in contact with media of the same refractive indices occurring in vivo. In *Limulus* the outer surface of the eye may be in contact with air or water since it spends its life partly in the sea and partly on the beach. In my experiments I thus chose air. The inner boundary, i.e. the marginal and apical surfaces of the cones impinge upon cells soaked in haemolymph and most of which are pigmented.

For similar investigations on insect eyes I had previously determined the refractive index of beetle haemolymph (from *Hydrophilus*) as 1.346; here a highly diluted glycerine solution served as an immersion fluid for the hind surface. As I had no reason to suspect that *Limulus* haemolymph had a significantly different refractive index, I used the same solution. This is all the more permissible as I found that the performance of the dioptric apparatus was remarkably unaffected, whether I set it up correctly as described, whether I immersed it completely in water or in very highly refractive fluids (e.g. aniline with $n = 1.5803$, i.e. higher than in normal glass). This proves that in *Limulus* practically all refraction occurs within the dioptric apparatus and that the surface boundaries play virtually no role.

Nevertheless, the quantitative experiments were performed with the dioptrics mounted correctly. The front surface of the eye, which had been freed from all adhering soft parts, was moistened with diluted glycerine so that it would not dry out and impair the image quality. This can be done with confidence, since a sufficiently thin film has no influence upon refracted or incident rays whatever its refractive index. The eye was then mounted upon a microscope slide with a hole drilled through it. The concave surface, facing upwards, was filled with a drop of the glycerine solution and finally a coverslip fragment was laid on top so that the emergent ray path was not influenced by the curvature of the fluid.

The preparation is then brought under the microscope, naturally without condenser and using only a plane mirror, or better still, after carefully fixing the eye, the microscope is turned on its side and directly aligned with the objects whose images are to be observed. This was my normal procedure.

Upon focusing on the cones, one is immediately struck by the strong convergence of their axes mentioned above. If the cone in the middle has its axis coinciding with that of the microscope, then even the six neighbouring cones deviate markedly from this direction, and at the edge of the field (one works naturally at low magnification) the cones already appear in half-profile.

The images of external objects lie near the apical surface. They are much larger than those usually seen in insect corneas and are not very sharp. If one focuses on the bars of the window one readily recognizes the poor resolution as the image shows very pronounced field curvature – the level of focus changing considerably from the centre of the apical surface to its edge.

It is thus naturally impossible to precisely pin-point the position of the image. Furthermore, there are small variations between different cones. In some cases I found that the image (in the centre of the apical surface) coincided exactly with the surface, in other cones the image appeared somewhat proximal to this. It was almost never seen to lie distally.

If the eye is aligned with a point source (a gas burner surrounded by a white clay cylinder and a black tin cylinder both having a ca. 1 cm diameter hole aligned with each other) then one naturally sees the bright image point in the apical surface of those cones which are more or less aligned with the microscope. No light emerges anywhere on the marginal surface of these cones or in the spaces between them. This demonstrates that all light incident approximately axially reaches the apical surfaces of the cones and that none emerges between them. This is no longer true for cones with axes deviating considerably from the line of sight. In these one sees focal lines on or near the marginal surfaces. The origin of these is easy to interpret: for rays incident perpendicular to its long axis a lens cylinder behaves as a cylinder lens, focusing the rays into a line. Obliquely lying cones are halfway to this situation. In fact one can often follow quite clearly how the light point of the cone in the centre is drawn out radially in the neighbouring cones, and gradually transforms into a focal line in peripherally located cones, lying first near the tip and then becoming longer always within or near the marginal surface.

If one makes a thick section through the cones, so that both front and back surfaces are removed, the remaining middle portion still functions as a convex lens, the inverted image that it produces simply lying at a greater distance than with undamaged cones.

From the foregoing it can be concluded that the *Limulus* cone (considered of course from the corneal surface to the apical surface) is a dioptric apparatus which functions essentially as a lens cylinder with a length approximately equal to its focal length. Since it is long in comparison with lens cylinders in other eyes, it produces a correspondingly larger image.

I determined the size of the image by using two point sources as objects and measuring the separation of their images. The magnitude of the image of a 22 cm object at a distance of 150 cm from the eye, is 0.043 mm.

If one focuses upon both images of two such point sources in the plane of the apical surface and then raises the focus (towards the back of the eye), then each image dissolves into a blurred circle. The centres of the two blurred circles move towards each other and merge into one. In one example the blurred circle then had a diameter of 0.13 mm.

I could also demonstrate the convergence of the emergent ray cones for shorter object distances. However, sometimes the convergence was less pronounced and sometimes the emergent cones appeared almost parallel. I never observed divergence. Since I never observed parallel emergence in completely fresh eyes, it is impossible to say whether one is dealing with in vivo or post mortem differences. Such small variations are of only minor physiological significance.

Thus with respect to the direction of the refracted principal rays (and consequently in all fundamental points) the *Limulus* cone has the optical function of a lens cylinder with length approximately equal to its focal length (see Chap. I). Where the principal rays show the convergence described, one must assume a somewhat greater length.

In order to obtain a clear idea of the retinal image over the whole eye, we must ask how large is that part of the visual field which delivers light to the apical surface of one crystalline cone, and hence, presumably, to the underlying retinula.

I found that the image of a light point crosses the apical surface when the light point is moved through 8.5 cm at a distance of 63 cm from the eye, a visual angle of 8°.[6]

On the same eye, the radius of curvature of the front surface, measured with the ophthalmometer, was 7.4 mm. Finally, I measured the separation (distally) of neighbouring cones. The result varied by ca. 10% with an average value of 0.28 mm.

Due partly to natural variation and partly to the difficulty of measurement, these values are only approximate, but they are sufficient to give an idea of the performance of the eye.

With a radius of curvature of 7.4 mm and an acceptance angle of 8° (which may be assumed for all cones) a cone separation of 1mm would almost suffice for light from all points on the object to reach the back of the eye. However, the cone separation is 0.28 mm (i.e. three or four cones per millimetre). From this it can be concluded that each point on the object is imaged simultaneously in several cones.

Direct observation confirms this. If a light point is placed so that its image lies in the middle of an axially aligned cone, the image can still be seen (admittedly lying markedly eccentrically) in the six neighbouring cones. Thus in a section through the eye, as the calculations suggest, three cones will all guide rays from an external point to their apical surfaces so that they should be able to strike the sensitive elements.

1. Function of Oblique Cones

Up to now it has been assumed that the axes of the chitinous cones are arranged radially on the corneal surface. However, as already mentioned, towards the periphery of the eye the cones become increasingly skewed. This has no significant influence on the properties already discussed, but it does lead to a broadening of the visual field of each eye, and since the *Limulus* eye is only weakly curved this could in fact be very useful. Because of their life-style it may be that these animals cannot afford to have a strongly curved eye. Thus they bury themselves in sand (cf. Brehm's Tierleben, 1st ed., VI p. 656), whereby the curved carapace, in which the completely immobile eyes are inserted, must push stony material to the side. In fact, in the specimens which I saw, the eyes were already scratched, and this would certainly have been worse if the eyes had stood out from the carapace due to a stronger curvature.

How this arrangement leads to a broadening of the visual field becomes clear when one considers the refraction of rays entering the dioptrics. In Text Fig. 9a central and a peripheral cone are indicated schematically. The central cone (*A*) focuses rays incident perpendicular to the cornea (*s, t*) to a point (*a*) on the cone axis near the apical surface. Rays incident with small angle (e.g. β) to the axis are focused in the same plane beside the axis (*b*) thus forming the inverted image *ab*. Rays incident with a greater inclination (α) will be absorbed by pigment (at *f*) and thus do not contribute to the image.

In the peripheral cone (*B*) the cone axis, *pq*, is no longer perpendicular to the corneal surface. A ray (*no*) incident perpendicular to the surface thus no longer reaches the apical surface but will be absorbed by pigment after following a path given approximately by *om'*. Even rays incident parallel to the cone axis, *pq*, will be absorbed by pigment after following the path *poq'*. Only rays like *oc*, which make a moderate angle to the cone axis and a substantial angle with the perpendicular, *nm*, travel parallel to the cone axis after entry, and thus behave like rays which in the central cone come from object points lying on-axis. *oc* represents the optical axis of the cone and should be distinguished from the anatomical axis, *poq*.

When, as in Müller's theory, each ommatidium of the compound eye is sensitive only to light incident approximately along its axis (which, as we shall see, is

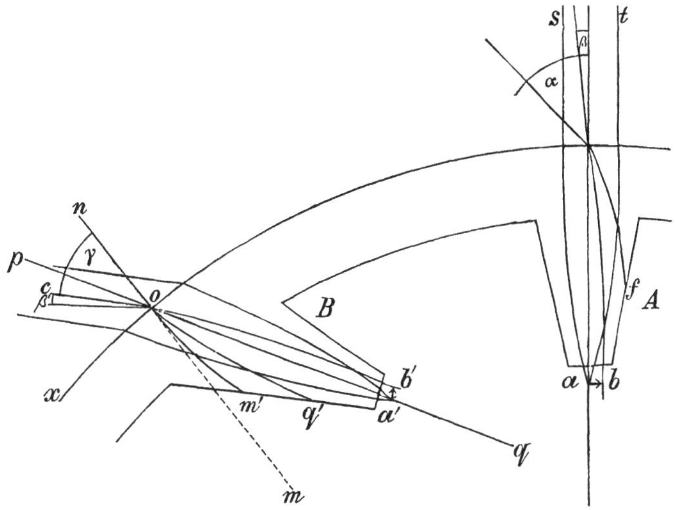

Text Fig. 9

not strictly true for many eyes) and if the ommatidial axes lie perpendicular to the corneal surface, then the degree of curvature of the cornea (or more precisely the fraction of a spherical surface represented by the cornea) is a measure for the size of the visual field. The arrangement in the *Limulus* eye enlarges the visual field even more than if the cones were to keep the same orientation and the degree of curvature of the cornea were increased so as to be everywhere perpendicular to the cones: indeed an elegant solution to the problem of producing a large visual field in an eye whose form is constrained by the need for protection from physical damage.[7]

The degree of the enlargement of the visual field can also be estimated. Thus the arc of the cornea in its longest (horizontal) meridian encompasses ca. 90°. The angle between the cone axis and the normal to the cornea can reach at least 40°. Furthermore, the refractive index of a piece from the most distal layers of the eye, measured with the microrefractometer, is between 1.5381 and 1.5327 – let us assume 1.535. From this the angle of incidence of a ray which should travel parallel to the cone axis after passing through the cornea is calculated to be 79–80°. Thus the visual field is extended by about 80° on each side and similarly ventrally (dorsally the overhanging carapace obscures the field). In the horizontal plane the field thus covers $2 \times 80° + 90° = 250°$. This seems almost excessively large, but is readily understandable when one considers that it will be considerably reduced as soon as the animal is under water.

This characteristic eye structure is also found in other animals (see Chap. IV). It would be worthwhile to follow it up in the different taxonomical classes and to examine the behaviour of the animals concerned.

2. *Function of the Cone Sheath*

As Grenacher has already described one can distinguish two parts of the cone from their characteristic refraction, the conical core and the surrounding sheath (best seen in Text Fig. 10, see also Fig. 20). With the microrefractometer one finds that the core has the structure of a lens cylinder with the lowest refractive index in its outer layers, and that the sheath is of higher refractive index. The indistinct boundary between them represents a rapid transition from weak to highly refracting layers. I have come across cones where there was also no doubt that the sheath itself was layered as well, but with refractive index increasing from the inner to the outer layers. However, the optical properties of different cones are not identical and the interpretation of the shades of colour seen in the microrefractometer with two lens cylinders of different types, one lying inside the other, is so difficult that I had to give up the attempt of clarifying this detail further. Nevertheless, there is no doubt that the core of the cone is a lens cylinder as described above and that its weakly refracting outer layer adjoins strongly refracting layers which form the inner layers of the sheath. (I have found similar complex structures in the cone and sheath of butterflies using the microrefractometer. Fig. 57, 2 and 3.)

The microrefractometer also shows that the nipple-shaped protrusion in the corneal substance (Text Fig. 10*a*) functions as a converging lens, i.e. the layer proximal to it is of lower refractive index.

I believe the significance of the optical separation of the cone core and the sheath is obvious and is elucidated in Text Fig. 10. *mn* is an axial ray, *pq* is the

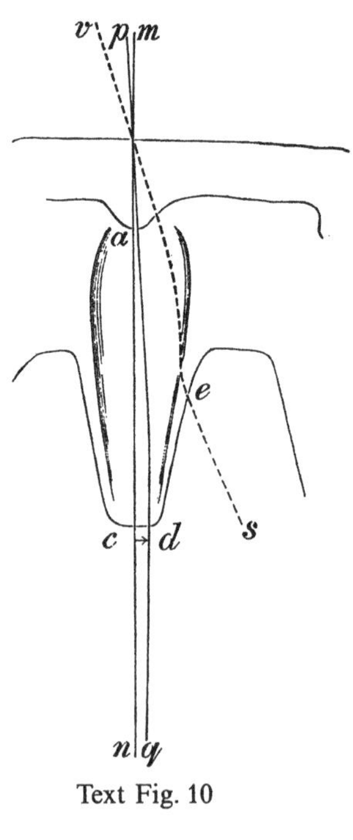

Text Fig. 10

course of a principal ray originating from a point near the axis, *cd* is the image. A ray, *vs*, coming from a more laterally situated point is not deflected strongly enough (by the curved surface at *a* and the lens cylinder) to reach the apical surface. Instead it passes through the boundary layers of the cone core into the adjoining, more highly refractive, layers of the sheath. Here the ray must diverge from the axis and ends up in the neighbouring pigment where it will be absorbed.

Two things are achieved by this arrangement:
(i) The angle of incidence with which such a deflected ray will strike the boundary of the cone at *e* is significantly reduced and thus the possibility of internal reflection is minimized – in fact probably completely avoided. This is important for the resolution of the image *cd*, since internally reflected rays would reach the apical surface, thus impairing the image quality.

One will now certainly ask why should *Limulus* have such an arrangement for the avoidance of internal reflection and why is the encasing of one cone in another not more widely found. Actually I am not sure that this formation is not of rather general occurrence and it is certainly present in some species. It is a fact, that in no eye have I found optical phenomena that would, with certainty, result in internal reflection.[8] On the contrary, the optical properties usually seemed to exclude internal reflection, at least for those rays which would degrade the image. Normally, such reflections are only to be expected; for example if one fashions the shape of a crystalline cone out of glass such reflections are abundant. Of course the glass does not have a lens cylinder construction, but even in crystalline cones one would, in general, predict internal reflection. This is why I have shown how this is avoided in *Limulus*. Another mechanism will be discussed later (Chap. II C).

(ii) The second consequence of this optical arrangement is that there is no transition between rays which pass through the apical surface without reaching the strongly refracting sheath layers and those which pass into the sheath. The former will be continually guided towards the cone axis, whilst the dioptric structure ensures that the latter will sooner or later follow a completely different direction, but never first in the vicinity of the apical surface. Under the appropriate conditions of illumination the harmful (off-axis) rays will therefore emerge all over the marginal surface of the cone except in the direct vicinity of the apical surface. The retinulae lie directly behind the apical surfaces and the elimination of rays which might strike the retinulae of neighbouring cones seems to be of functional importance, despite the presence of screening pigment.

This arrangement results in very beautiful and conspicuous images. When the apical surfaces are viewed in the microscope with appropriate illumination (from a window) the horizontally aligned apical surfaces appear bright whilst their surroundings are dark – as already described. In other cones, lying somewhat obliquely, not only the apical surface but also most of the margin appears bright. Only the most proximal part of the margin bordering the apical surface appears dark. In fact this band gives the impression that there is still pigment adhering to it and the preparation must be further tilted in order to see that this is not the case. This dark belt is the demonstration of the optical arrangement described.

3. The Combined Function of the Cones and Retina

Following this explanation of the optical function of individual crystalline cones, we can now consider how they combine to produce the retinal image. For this, it is above all necessary to know the physical location and extent of the retinal elements, and also whether the retinula functions as a single unit producing only a local signal, or as a retina in the sense of a vertebrate retina. The latter I believe to be false but I will return to this question later. For the present we shall assume the retinula to be a single visual unit.

Unpigmented cells lying on the apical surfaces form a direct continuation of the cones (Fig. 20). Grenacher was not sure whether these represented processes of the underlying retinula cells or not.[9] This unpigmented cell bundle tapers proximally, making way for the rod formations (Grenacher's rhabdoms); in cross-sections these form delicate stars of very small size (Fig. 21). The whole unpigmented mass (i.e. retinula plus cell bundle) is cone-shaped, the proximal part contains the rhabdom and is ensheathed in dense pigment which is almost continuous with that surrounding the chitinous cone. According to Grenacher, part of this pigment is contained in the retinula cells themselves, the axial surfaces of which actually form the rhabdom. Grenacher also describes nerve fibres merging directly with the retinula cells.

If the unpigmented cells adjoining the apical surface are light-sensitive they will be hit by all rays that originate from a point on the cone's optical axis and which are incident on the overlying facet. These form a cone of light with the corneal facet as base and the object point as apex. Without the light-collecting dioptrics, far fewer rays would reach the retinula from each point, as the retinula would then form the base of this cone. Further, the retinula will also receive light from equally large light cones originating from all points within a visual angle of 4° from the first one. The principal rays from these light cones leave the apical surface parallel to the cone axis or converge towards it. If one or more spherical refracting surfaces had been used for light collection, the principal rays would diverge, i.e. there would be a much greater variation in the directions of all the rays. If the degree of nervous excitation depends upon the path length in the light-sensitive cell, or if the light-sensitive layer does not lie directly on the apical surface but further behind it, then it is clear that the lens cylinder is superior to the spherical surface.[10]

However, it is much more likely that light sensation first occurs in the rhabdoms. Grenacher has shown that these are the most conservative elements in the eyes of all animals, and we know that in the human eye the rods and cones coincide closely with the light-sensitive layers.[11] If the rhabdom is the mediator of light sensitivity then a parallel organization of the principal rays is of obvious advantage. One can further see that not all points belonging to the elementary visual field (to use Notthaft's terms) direct the same amount of light to the rhabdom. Towards the periphery less and less light from each point will be perceived, since a part of the associated blur circle will be lost for perception.[12] If the principal rays converge behind the apical surface (as I sometimes saw) then they will be directed towards the rhabdom together with most of the rays of the associated blur circles. Here the advantage of a lens cylinder is even more apparent. The axial point will still produce the maximal excitation – unless the sensitive layers are infinitely thin and lie exactly where the blur circles from all points in the elementary visual field coincide, or with thicker layers the plane of coincidence must lie exactly in the middle. Even then a strict correspondence of all points in the elementary visual field would not be maintained, since the rhabdom lies in a narrow channel surrounded by pigment.

Brücke was the first to point out how light is captured in the analogous structures of the vertebrate retina, the rods.[13] Because of its high refractive index (over that of the surroundings), light that enters the rhabdom with less than the critical angle will be guided by internal reflection to its end, where it can be at least partially reflected and guided back in the same way. Whatever the origin of nervous excitation, this entrance of light into the rod formations (also represented by mammalian cones) appears to be of fundamental importance. It is sufficient for an eye to be so constructed that light enters the rhabdom only at one end even when, as in *Limulus*, most of the rhabdom is embedded in pigment.

From all this it follows that we must imagine the perceived retinal image in *Limulus* to be erect and with a resolution such that a grating with 13 cm bars will still be recognized at a distance of 1 metre, although the edges of the bars will not appear sharp. The brightness of the retinal image compared with that of the object – "the relative brightness of the retinal image" – is quite high, in that each retinal element receives more light than in most other arthropods. Correspondingly, the number of retinal elements per unit area is very small. The "relative size" of the retinal image is also large. Taking the radius of curvature of the eye as 7.4 mm and the distance of the sensitive surfaces from the corneal front surface as 0.92 mm (approximate values taken from data and drawings of Grenacher) one can calculate the length of the retinal image of a 1-metre object at a distance of 1 metre as 6.5 mm. In the human eye the retinal image of the same object would be 15 mm. The extent of the retinal image, or visual field, is appreciably larger than the curvature of the eye would suggest and as we have seen covers a horizontal angle of 250° when the animal is in air.

Can one see the retinal image? According to the theory set forth this must be possible, even if imperfectly. Experiment confirms this. First, the surface of the eye preparation is moistened with a little glycerine dissolved in alcohol in order to compensate for the roughness of the cornea resulting from damage. If one then aligns the eye with a window and observes it from behind, initially one sees no clear image. However, with the aid of a magnifying glass a larger group of apical surfaces appear bright whilst the spaces between seem dark. These are the apical surfaces of the cones with axes directed towards the window. The bright spot formed by all of them together is the upright retinal image. It moves in the corresponding direction when one rotates the eye, and an object moved between the eye and the window leads to a darkening of rows of apical surfaces in the way expected for an upright image.

The image is not perfect; first because one sees the entire apical surface as bright and not (as appropriate to the function of the eye) the much smaller surface corresponding to the cross-section of the rhabdom; and secondly because the area surrounding the image of the window does not appear dark. As discussed above, this is because light rays entering with a rather large angle to the optical axis emerge from the marginal surface. Normally these would be absorbed by pigment, but in the eye preparation they brighten the whole surroundings of the real retinal image with the exception of a narrow zone around the retinal image, which is also explained in the above account. If one ignores all the light streaks leaving marginal surfaces as physiologically irrelevant, one can study the retinal image and its movements with ease. I have also quite successfully managed to replace the pigment with a black lacquer which is absorbed into the spaces between the cones by capillary action leaving the apical surfaces free. The clarity of the retinal image is considerably improved as a result.

I describe below how the pigment in insects migrates under the influence of light, occupying a different position depending on whether the animal is in the light or dark. This may also be the case in *Limulus*; but since I have made no relevant observations, I can only speculate. It is possible that, on illumination, the pigment sheath around the retinula changes its width, or varies its distance from the apical surface as a result of a reduction in the width and/or height of the unpigmented cone which adjoins the apical surface. A narrowing of the pigment sheath would correspond to the light-adapted state, a broadening would increase the brightness of the retinal image, but probably at the cost of its sharpness.[14]

4. *The Eye of Limulus Compared to Those of Trilobites*

Of all the eyes of living insects and crustaceans with which I am familiar, none is more compatible with Gottsche's hypothesis than that of *Limulus*, but even this offers only weak support. If one were to assume that the transparent cells adjoining the apical surface were the retinal elements, if one considers the small number of nerve fibres which Grenacher saw making connections with the retinula cells as sufficient for monitoring an inverted image of the outside world in every ommatidium, if one wants to ignore the futility, or even disadvantage, of using lens cylinders instead of spherical refracting media, and also the many further criticisms raised against this theory, then one could consider the inverted image lying in or behind the apical surface, imperfect as it is, as the retinal image, the transparent cells as the retina, and say that *Limulus* sees with numerous eyes, each of which functions in the fashion of a vertebrate eye. I believe that Schultze would have rejoiced to have seen the inverted image and the relatively numerous retinal elements (Grenacher estimates 14–16), but I have a different opinion as to how the eye functions, even though I have just emphasized the convergence with the vertebrate type eye and can provide another link between this and *Limulus*.

The taxonomical position of *Limulus* is unclear; earlier it was considered close to the fossil trilobites but this has been questioned recently.[15] Professor Suess was kind enough to let me have a pair of trilobites, whose eyes, with their large easily visible facets, aroused my keen interest. In this long-extinct relative of *Limulus* I believed to have rediscovered its eye, only on a much larger scale.

Using the method of mineralogists, I ground thin sections from the eyes, some perpendicular and some tangential to the surface. In the microscope the tangential sections already appeared very different from *Limulus* (Fig. 18) with regularly spaced circular fields separated by septa. But one could hardly consider the septa as being embedded in a coherent chitinous substance. The circular fields might have represented cross-sections of unusual cones, particularly as they showed the same regularity in their spacing as the facets. However, sections perpendicular to the surface showed no signs of cones but exhibited very beautiful regular lenses similar to the chitinous formations in insect ocelli. A pair of these from *Phakops fekundus* is shown in Fig. 19. Unfortunately, I could recognize nothing more from the deeper parts of the fossil eyes. Yet, what I did see appears to show that the optical principle of the lens is in operation, and the shallow ending of the septa between the lenses leads one to suppose that the retina lay close behind the lenses. In a word, we are looking at a simple eye, and such an eye undoubtedly sees with an inverted retinal image.[16] Paired simple eyes are well known from spiders; in trilobites there are not two, but many together forming a complete eye with a very small curvature – less than in *Limulus*. If one can draw a conclusion from these meagre findings, these animals thus really saw in the way demanded by the theory of Gottsche, Gruel and their followers: many aggregated eyes, each projecting a small inverted image, worked together as the visual organ, and if one only took into account those rays which strike the centre of each retina, a large erect image would result in each complete eye. Even the gross difference between these eyes and that of *Limulus* is evidence for the change in functional strategy, the inefficiency of the small inverted image and the perfection of the large erect one.

B. The Superposition Image of Lampyris

In the upright image of the *Limulus* eye, light entering neighbouring facets is focused at the level of the retina in adjacent positions. I have attempted to indicate this with the name apposition image. In the next example, which serves to illustrate the second type of compound eye, the light entering neighbouring facets is largely superimposed at the level of the retina. I thus call this type of retinal image the superposition image.[17] For obvious reasons I will, initially, again restrict myself to the eye of one animal. This is the eye of the firefly (*Lampyris splendidula*) which, as in *Limulus*, is amenable to investigation since the entire dioptric apparatus is in one piece. I only used the eyes of the flying males, as those of the wingless females are too rudimentary.

1. Observations on the Fresh Lampyris Eye

After slicing off most of the nearly hemispherical eye with a sharp cataract needle, I placed it in a small dish and brushed out the concave side whilst holding the eye steady with a needle or fine forceps. In fresh eyes the pigment comes away easily, though eyes preserved in alcohol present some difficulties. A drop of diluted glycerine with refractive index $n = 1.346$ is now placed on a cover slip or a flake of mica. I chose this fluid to simulate, as closely as possible, the natural situation, where the crystalline cones are in contact with tissue fluid. The excised eye is placed with the concave side lying on the drop, whilst the convex side remains dry and in contact with air. This is, in fact, simple because the fresh corneal surface is hydrophobic and the eye ends up in the desired position almost automatically. I prefer mica to a normal coverslip as the drop spreads out better.

As usual, the mica (or coverslip) is then laid over a slide with a hole of about 1 cm diameter, the eye of course being suspended in the opening, and the slide brought under the microscope. As in the living animal, the front corneal surface is now in contact with air, and the crystalline cones with a liquid of $n = 1.346$. By raising the focus, preferably using a low magnification of 60–100 ×, one sees an upright image of external objects (disregarding the inversion due to the microscope and the effect of the microscope mirror). An unretouched microphotograph of such an image accompanies this text as the frontispiece. The eye faced a bay window through which a church was visible; on one of the panes I had stuck the letter R cut out of black paper.

To rule out deceptions due to the effect of the concave mirror or other reflections and refractions, one can use a plane mirror, or one can place the object, e.g. a cataract needle, between the mirror and the preparation, or replace the mirror with paper; one can remove the mirror, lay the microscope on its side, pointing at the window, and move an object around in front of the preparation. Under all these conditions one can still recognize the image as upright with a simple microscope or Brücke's magnifier. Indeed, I have no doubt that a very short-sighted person could see it without any optical aids at all. I mention all this because probably anyone, myself included, on seeing the image for the first time would think that an extraneous image might somehow have resulted through double reflection from the microscope lenses or the preparation itself. In the following, any remaining doubts should be completely dispelled.

Sometimes I also used well slides, but later abandoned the practice, firstly because the depression functions as a convex lens, thus introducing a refraction which must be avoided in more exact investigations of the optical properties (for this reason I also worked with a plane mirror or none at all), and secondly because evaporated water condensed in the well, and soon blurred or destroyed the image.

Conversely, observations in free air, as just described, have the disadvantage that the refractive index of the fluid changes, and this cannot be tolerated in more exact tests of optical behaviour. In order to prevent errors one must then constantly renew the drop. Incidentally, the beauty of the image is only slightly dependent on the exact value of the refractive index. Remarkably, with water it is just as good; however, concentrated glycerine should not be used.[18]

The definition of the image exceeded my expectations. In a speculative illustration, Forel (1886) shows the definition of the retinal image that a bee might receive from a small insect as it flies by. From my earlier investigations I had also expected images of approximately this quality. However, in *Lampyris* they are sharper, and there is every reason to assume that the eyes of other insects are even more perfectly constructed. I could discern the shape of a cataract needle held between the preparation and the mirror very well, and could recognize the white handle and the reflection as the needle was turned. I made some of these studies in the country during the summer holidays, and there the image of the microscope mirror served as a frame for a miniature landscape in which I could discern the white brick pillars of a barn lying opposite my window, its red tiled roof and brown hoardings, and in which single fine branches of small damson trees stood out against the blue sky. I will discuss the definition of the image further at a later stage.

Because of the somewhat vague edges the size of the image cannot be measured accurately. In one case the length was ca. 0.24 mm for an object 32 cm long at a distance of 52 cm.

Before discussing the position of the image in the depth of the eye, i.e. its distance from the corneal surface, it is necessary to describe the most important aspects of the eye's anatomy.

Figure 1 shows a meridional section through the *Lampyris* eye. For each crystalline cone, the convex front corneal surface bears a curved facet with radius of curvature greater in the centre and smaller in the periphery of the eye, varying from 0.09 mm to 0.02 mm. Assuming the cornea to extend to where the chitinous substance separates into single cones, then in this animal it is very thin, and in a section of an unprepared eye can be recognized by its lack of pigment. The crystalline cones (*k*), which are fused with the cornea, are thickly ensheathed in pigment (*J.P.*) except for their proximal ends which project freely into the underlying cells. The shape of the cones can only be recognized in eyes which have been brushed out. In Text Fig. 12 I have reproduced the shape of a cone and its corneal facet as faithfully as possible. There now follows a rather broad zone of elongated cells (*G*) containing nuclei and arranged radially. These are probably identical with similar structures in lepidopterans, and their significance will be discussed later. The retina begins at *R* , but one can see very little of this in meridional sections. In horizontal sections one recognizes a cell layer rich in nuclei. Towards the retinal pigment layer (*R.P.*) its mosaic changes abruptly into an extremely elegant form composed of forget-me-not-like shapes. The mosaic can still be recognized, though in modified form, both within and proximal to the retinal pigment layer. The centre of each flower-shaped cross-section is unstained; this corresponds to Grenacher's rhabdom. Deeper still one can see the thin layer (*m.f.*) into which the nerve fibres from the optic ganglion (*G.o.*) project and disappear.

The position of the image is very difficult to measure precisely. The position where the upright retinal image has its maximal resolution can be determined relatively accurately; the problem lies with the second point, namely, the corneal surface. A position can be accurately found where each facet appears as a bright circle and the spaces between them dark, even after the pigment has been brushed away. As I explained in more detail for the eye of *Limulus*, this is because refraction prevents light from penetrating the transparent substance between the facets. However, I cannot say with certainty what level this position corresponds to; possibly the base of the crystalline cone, perhaps the base of the curved facet, but maybe also some more proximal level in the crystalline cone.

By focusing further back on some remaining pigment granules I believed I could focus accurately on the apex of the crystalline cones. The distance from this position to the retinal image I determined as 0.23 mm.

If we were to transfer this image to the drawing (Fig. 1), it would fall appreciably behind the retina. This discrepancy can possibly be explained in that my drawing represents a meridional section from the lateral eye region which I chose because I could depict the entire section without making the drawing too large.

On sections through the centre of the eye one sees that not only does the curvature of the facets increase towards the periphery, but also the cones become shorter (e.g. I measured 0.055 mm compared to 0.082 mm in the centre). Thus the whole dioptric apparatus becomes stronger and the image lies nearer.[19]

Another possible reason for the discrepancy is shrinkage. For example it is well known that white blood corpuscles in blood vessels fixed in alcohol appear smaller than fresh ones. It is quite possible that, despite the supporting chitin exoskeleton, some reduction in volume also occurs during the preparation of the eye for sectioning.

Without these uncertainties it would theoretically be possible to make a more accurate estimate of the position of the sensitive layer within the retina from the position of the image. One day this may be possible.

Although of no physiological importance, the following optical curiosity is of some theoretical significance. An image can also be seen when the eye cup is viewed from the other side (i.e. with the glycerine-coated concave side facing the object, and the convex side, in contact with air, pointing towards the microscope). Remarkably, it lies in the same position as the normal retinal image, i.e. proximal to the optical apparatus; it is of the same size, but inverted.

2. *Overview of the Dioptrics of the Lampyris Eye*

In this section I shall present my view of the dioptrics of the *Lampyris* eye, subsequently I shall present evidence for my conclusions and also describe how they were reached.

If Müller's theory in its original form were correct for the *Lampyris* eye, one should be able to see the erect image of an object when focusing the microscope on the tips of the crystalline cones. Although one sees a suggestion of this, a much clearer image is found appreciably behind the tips of the crystalline cones. With certain assumptions this might still be compatible with the theory, but the result of the following experiment is completely incompatible.

A *Lampyris* eye cup is prepared as described and placed on the stage of the microscope. The microscope is now aligned with a point halfway between two point light sources (e.g. two candle flames) without intervening refracting or reflecting media. If one focuses at the level of the retinal image one naturally sees two light points – the images of the candle flames. If one focuses towards the cornea one sees the optical cross-sections of the ray bundles which give rise to the two image points. A group of rays converges on each point, each ray coming from one crystalline cone. If the two candles are appropriately spaced, then two rays can be seen emerging from the majority of the illuminated cones, one in the direction of each image point. A ray entering the cone from the right hand object is deflected towards the right hand image point, and a ray entering the same cone from the left hand object towards the left image. In other words: a ray entering a cone with a certain angle to the cone's axis leaves the cone with the same angle; the axis, incident and emergent rays lie in the same plane, with the emergent ray lying on the same side of the axis as the incident.

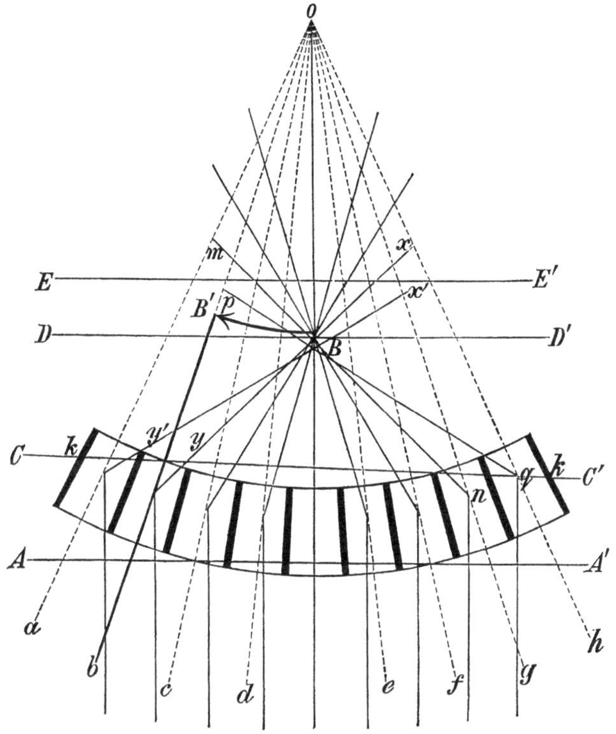

Text Fig. 11

This would seem to suggest that there is a reflection within the crystalline cone; however, the effect is in fact a result of refraction. Before discussing this further, I will attempt to clarify the foregoing with the aid of diagrams.

Text Figure 11 illustrates the paths of rays leaving an object at such a distance that the rays can be considered parallel on reaching the eye. (Because of the small radius of curvature of the eye this is true even for short distances). The figure corresponds to the situation under microscopic investigation with the stage horizontal. *kk* represent the individual facets, *oa* to *oh* their axes, the continuous vertical lines the incident rays, which are deflected as shown to converge at the image point *B*. *o* is the centre of curvature of the eye. One can see that a completely analogous image *B'* must result from a second object lying along the extension of *ob*, and that at least some of the same crystalline cones are involved in its formation. In addition one can see that the image is upright, what determines its size, etc.

If one focuses the microscope at the level *AA'*, which may lie in front of the cornea, tangential to the cornea or perhaps even in the distal portions of the crystalline cones (as already emphasized it is not possible to determine this accurately), one sees the image reproduced in Fig. 2. This picture, described already in my first paper,* represents the the optical cross-sections of the rays leaving the

* However, at that time I considered the level at which it was seen to be somewhat more proximal.

light source in the focal plane of the microscope. These are, of course, not the incident rays themselves, but the continuations of the rays, *pq, mn* etc. which eventually form the image.

On focusing up and down the image remains basically the same over a relatively large distance. Only the change in the distance between individual light points shows that the rays are actually converging. If one uses two light points as the object then each point appears double. Figure 2 also shows indications of diffraction phenomena on most of the points, as would be expected under the prevailing conditions.

On focusing further up, towards the region of the cone tips (plane *CC'* Text Fig. 11) the image of Fig. 2 gradually transforms into that of Fig. 3. The bright points become circles, most of which have a particularly bright centre and one or two more or less distinct haloes. Here again interference phenomena are apparent, only a suggestion of this being indicated in the drawing. At this level of focus one sees no doubling of the circles when using two light points as object.

In my opinion the bright centre of each of these round spots contains the physiologically relevant rays, and the bright haloes are formed by rays normally absorbed by pigment, if in fact they are normally so abundant. Thus by changing the focus, one sees that the light in the haloes scatters diffusely in all directions. Further I have found that the haloes are much less pronounced in completely fresh preparations, so that Fig. 3 becomes very similar to Fig. 2, the light points simply being closer together.

Raising the focus still further, one encounters rays between the dioptric apparatus and the image point. The circles of Fig. 3 draw closer together and form intricate interference patterns. With two light points one sees that the circles are again double. As one focuses towards *DD'* one gradually approaches the image point, represented in Fig. 4 . Corresponding to the regular hexagonal array of the crystalline cones, one sees also here diffraction phenomena reminiscent of the striations in *Pleurosyma angulatum*. Finally, if one focuses on a plane, *EE'*, behind the image point, it diverges into a blur circle (Fig. 5), in which one recognizes the interference pattern of a three-pointed star as well as other indications of diffraction spectra.

I should mention that Figs. 2–5 all come from one eye prepared in the normal manner and viewing a single point source; in each drawing the meridian had the same orientation with respect to the hexagonal array of crystalline cones.

Thus, the *Lampyris* eye forms an erect image by means of convergence as in the vertebrate eye; if we define the axis of the eye as the line between any point on the object and its image, then, as in the vertebrate, rays parallel to the axis are more strongly deflected the further away they are from the axis (beyond a certain point they are no longer involved in image formation). Further, the refracted ray remains in the plane defined by the incident ray and the cone axis etc. This is all achieved by a dioptric apparatus composed of crystalline cones fused with the cornea arranged radially on a spherical surface.

What dioptric construction must the crystalline cone have in order to achieve this function?

Any pair of convex lenses separated by the sum of their focal lengths and lying on the same axis performs basically the same function as the crystalline cone.

Oblique, parallel rays striking the first lens will leave the second lens with an angle to the axis which increases with the angle of incidence. Further, they lie in the plane defined by the axis and the incident ray, and are parallel with one another. Under normal conditions therefore, they could not form an image at a finite distance. However, if one considers the parallel ray bundle to have a microscopically small cross-section, one realizes that the convergence of a large number of them can result in an image point, since the other dimensions of the image are then large compared to this cross-section.

Doubtlessly the situation with the crystalline cone is not so simple as with this lens combination. Initially I believed that the corneal facet played the role of the first lens and the curvature of the cone tip the second. But an exhaustive study proved that this is not the case. At least to a great extent the function of the crystalline cone depends rather on the layering which characterizes it as a lens cylinder. Of course, the curved surfaces will complement this function. In this case the lens cylinder has a length equal to the sum of its focal lengths.

Text Figure 12 shows a close approximation to the path followed by two ray bundles in one ommatidium, drawn from observations on fresh eyes. *ab* is an object at a great distance so that the rays from it can be considered parallel. *m* and *n* are the boundary rays entering the facet from the base of the arrow (*a*). Using the measured facet curvature and the refractive index value I had determined for the cornea of *Hydrophilus**, refraction at the curved surfaces alone would result in rays converging to a single image appreciably behind the crystalline cone (see dotted lines m_2 and n_2). However, following the initially weak refraction the rays converge further as a result of lens cylinder action to form an image at a'. Similarly, rays *p* and *q*, from point *b* of the arrow will converge at b'. The image $a'b'$ cannot be seen as such, since the proximal portion of the crystalline cone still lies between it and the microscope. If a' did not lie exactly at the focal point of the proximal cone portion, but somewhat proximally, then the rays m_1 and n_1 would diverge slightly on leaving the cone and the virtual image mentioned above would be accessible to observation. I cannot pin-point the position of the image $a'b'$ exactly, and the drawing should only indicate that it lies in the proximal half of the cone. As I will later describe in detail, this image can also be made directly accessible to observation by cutting off the tip of the crystalline cone. Obviously one must see the image directly if one truncates the cone at the level $a'b'$ and views it from behind.

From the drawing one sees that an oblique ray is deflected towards the same side of the axis from which it arrived. One also sees that the angle it makes with the axis after refraction increases with the angle it subtended before refraction. I will demonstrate later that the ray bundles from a single point which pass through different cones are combined again into a single point. One sees further the significance of using not only two curved surfaces but also the lens cylinder principle to fashion the dioptric apparati, each of which represents a small astronomical telescope focused at infinity. If the curvatures of the corneal facet and the tip of the crystalline cone were sufficiently strong, the cone depicted in Text

* The cornea of *Lampyris* is so thin that I could not determine its refractive index accurately enough.

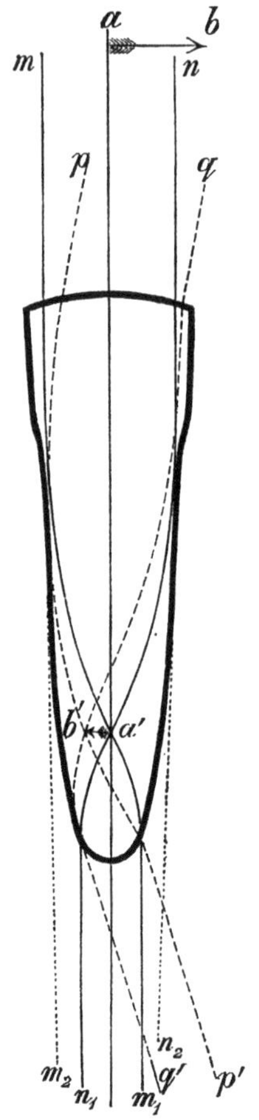

Text Fig. 12

Fig. 12 could function in essentially the same manner; however, one sees at once that rays incident with large angles to the axis, such as p, can no longer be involved in image formation, unless the refractive index has an enormous value.

In short, the dioptric apparatus functions as a lens cylinder with length the sum of its focal lengths (see Chap. I).

I have made a model of the insect eye along these principles which clarifies the optical function and which I used for the experimental proof of the formulae concerning the size, position etc. of the image which will be derived later.* It consists of ten pairs of convex lenses (spectacle glasses). Each lens has a focal length of two inches and is fixed on a board at a distance of four inches from its partner. Each pair represents the dioptric apparatus of one ommatidium. The ten

* Such a model, in a modified form can be bought at Lenoir and Forster, Waagasse 5, Vienna IV.

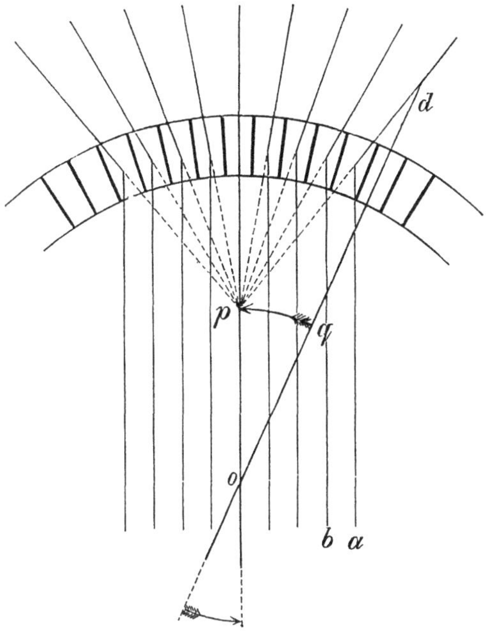

Text Fig. 13

pairs are arranged in an arc with 75 cm radius (measured to the common focal point of each pair). The model projects erect images, admittedly with poor resolution, since the ray bundle leaving each lens has the cross-section of the lens surface. Some five lens pairs contribute to each point of the image. Using the model, one can demonstrate, for example, the change in size and position of the image for different object distances. One can improve the resolution of the image by placing diaphragms in front of the lenses, though the number of lens pairs involved in the formation of an image point is reduced.

Finally, I return to the inverted virtual image which one sees when the eye cup preparation is viewed from the corneal side with the concave side facing the object. As already mentioned, this image lies in the position of the normal retinal image and finds a ready explanation directly from Text Fig. 13. All parallel rays incident on the crystalline cones will be deflected as shown to emerge divergently, and their extrapolated paths coincide at one point, *p*. Rays from another point on the object and incident on the eye with a different direction, e.g. *od*, coincide at another point (*q*). The two pairs of points, together with the common centre of curvature of the eye form two congruent triangles with opposite angle.

If one again uses a distant light source, which thus projects its image in the vicinity of the retina, and then focuses the microscope (now on the corneal side) proximally to the retina, one again sees the three-pointed star of Fig. 5. On focusing up the virtual image of the point source appears (Fig. 4), then the image of Fig. 3 and finally the optical cross-section through the ray bundles in the single cones as they are portrayed in Fig. 2. One thus sees the same images as in the normal direction, and, as far as can be judged, in the same positions.

3. Experimental Verification of the Ray Path in the Eye of Lampyris

I present now the observations which led me to the foregoing interpretation of the optical mechanisms in the *Lampyris* eye. Unless otherwise stated, it is assumed that the eye cup preparation, with the front surface in contact with air, and the hind surface immersed in glycerine solution ($n = 1.346$), is aligned directly with the object, so that light from the object reaches the cornea through air without undergoing any reflection. The image was observed through a coverslip or mica flake. In other words conditions are produced which approach the physiological situation as closely as possible.

If one focuses on the image of a bright point and introduces a mask from one side in front of the eye, the image disappears, all parts of it becoming simultaneously darker. If one focuses before the image point (corresponding to Figs. 2 and 3), then the bright spots corresponding to each facet disappear first on the side from which the mask is introduced. However, if one focuses behind the image (Fig. 5), the blur circles disappear first on the opposite side. (I describe here the actual behaviour, disregarding the inversion of the image due to the microscope). Thus in these respects the image and blur circle behave exactly as with a convex lens, except that in the insect eye one can see the individual rays as they converge to form the image. Normally this is represented with a ray diagram or a model with threads, here one sees it directly.

From this, as well as from the direct observations at different levels of focus described above, it can be concluded that the image point does indeed arise from the convergence of rays directed through a group of crystalline cones onto one point.

If one moves the object point the image also moves, from which it follows that the point at which rays from different crystalline cones intersect is not at the centre of curvature of the eye, as would have to be the case with Müller's original theory. If one focuses near the tips of the cones (Fig. 3), the individual spots of light are successively extinguished as the object is moved. However, the spots do not move at this level of focus. On focusing down (towards the object) the light points move in the opposite direction to the object, but if one focuses towards the image plane (so that an image between Figs. 3 and 4 becomes visible) the light points move in the same direction as the object.

Initially, I concluded from this that there was a form of total internal reflection at or near the level of Fig. 3, as I stated in my previous paper. However, this is not the case and we must make the distinction familiar from the two forms of mirages. A ray from a point on the earth can be reflected by the upper layers of the atmosphere. An observer then sees the object higher than it really is and inverted. The ray can also be refracted in a curve due to the refractive index of air layers decreasing with altitude. In this case the observer again sees the object lying higher, but now upright.*

* cf. Müller (1875), pp. 372ff. For an observer to see the mirage of an object lying in the same horizontal plane, Müller believed the air layers would have to be convex. From what we now know about the ray path in lens cylinders and similarly layered structures, this assumption becomes superfluous.

If one now uses two light points as the object, then at the level of Fig. 3 one sees the same image as with one light point. If one focuses down, then each circle of light in Fig. 3 separates into two bright spots corresponding to the cross-sections of rays diverging in the forward direction. If one masks the left hand object point the right hand bright spot disappears. If one now focuses towards the image point of Fig. 4, one again sees the separation into two rays, each directed at its image point. But if one covers the left object point, the left ray now disappears.

From this it can be concluded that we are dealing with a phenomenon of refraction and not of reflection. The same can be well demonstrated in my model of the *Lampyris* eye by placing a mirror between the two lenses of each "facet", parallel to, but displaced from the axis, and then observing the light bundle leaving the lens pair with and without the mirror.

The next point to be discussed relates to my findings on the optical construction of the individual crystalline cones and attached cornea.

If the eye cup is placed in a drop of fluid so that it is surrounded by fluid on all sides and observed under the microscope in the normal manner, one cannot see the erect retinal image. This is not surprising since the refraction at both surfaces is now altered and the incident ray bundle no longer leaves the individual cones with approximately parallel rays. However, one can see excellent inverted images under each facet which are not seen under normal conditions.

With a weakly refracting fluid these images might still be considered to originate from the end surfaces. However, if the eye is placed in aniline ($n = 1.588$), then a convergent function of the end surfaces is certainly excluded. Thus I had previously estimated the refractive index of the cone in *Hydrophilus* at 1.559, and, if this animal also has a layered crystalline cone, even this value is too large, so that in *Lampyris* refraction at the end surfaces is certainly neutralized in aniline. Nevertheless, each facet still projects its own inverted image. If the eye is mounted on a normal slide with the cone tips facing upwards and covered by a raised coverslip, the image is seen above the plane corresponding to the tips of the crystalline cones. Thus each facet has an image-forming capability independent of the end surfaces which can presumably only be explained by a layered construction.

If, under the same conditions, one focuses distally towards the cornea, to one's surprise one again sees an image, apparently one for each facet. The image, which is now erect, is much less obvious than the former, yet its existence cannot be doubted, especially if one introduces a moving object between the eye and the microscope mirror. Since it has no physiological significance it is not of great importance to know how it is derived, but the explanation is probably as follows. Under the microscope one sees the optical cross-section of the crystalline cone as a small circle, and on focusing somewhat deeper, that of the corneal facet as a larger concentric circle. Between the circumferences of the two circles lies a zone whose appearance does not indicate a high refractive index. This represents the marginal surface of the broader part of the cone. From its shape, and because it is surrounded by more highly refracting media, it could function as a diverging lens and thus produce the distal erect image, whilst the crystalline cone proper projects the proximal, inverted image as a converging lens. At least it appears, on focusing of the microscope, that the first traces of the erect image were to be seen in this zone whilst the inverted image appears in the inner circle. It is also possible that the sites near the corneal surface, where three facets adjoin, function as concave lenses. Even here the highest refractive index is found in the axis of the facet so that the refractive index gradually increases from the centre of such an interstice (though not in cylindrical layers). Because of the vagueness of the image and its position under the cornea it is not possible to fix its position accurately with respect to the single facets. Of the two explanations I consider the former more probable.

The following observations also lead to the conclusion that the inverted image depicted in Text Fig. 12 lies within the crystalline cone. An eye was embedded in celloidin, and serial sections cut tangentially to the cornea. Some of the sections of the eye are approximately circular, and in the centre one sees only the proximal truncated ends of the crystalline cones. Towards the edge, more and more of the distal part of the cone is present; with appropriate section thickness the proximal part of the cone is truncated whilst the cornea is still intact. Thus at the edge of the section only the most distal parts of the dioptric apparatuses can be discerned as round discs.

In the section I chose to examine, no complete cone was present since the thickness of the section was less than the length of the cone. Further, since the eye was embedded in celloidin ($n = 1.561 - 1.565$), refraction at the boundary surfaces could be ignored, at least for a converging function. (This was probably already replaced by a diverging function).

When the section was directed at two flames (85 mm apart and 585 mm from the eye), each facet projected an inverted image. Though I could not measure the distance of the images from the facets, at the edge of the section the image size (i.e. the separation of the two light points) was ca. 0.031 mm. Here the images lie significantly behind the facets and are formed only by the most distal part of the dioptrics (cornea). Towards the centre of the section the images decrease rapidly in size; I measured: 0.011, 0.004 and 0.002 mm. Naturally the images also shift closer and closer to the facets, and the last image, which was five times smaller than the first, lay – as far as can be judged – in the cut surface of a clearly truncated crystalline cone. (It is easier to convince oneself of this location using daylight and by observing the image of a window frame).

Further towards the centre of the section the cones appear only as small circles since only the proximal ends remain. Their images again increase in size, e.g. 0.005 mm and lie further away from the refracting medium. Thus, even without any refraction at spherical surfaces, it is possible to see an image in the section of a cone. The effect of the distal corneal surface must shift the image further distally. I cannot pin-point the position of the image in the cone more accurately; however, from what I have described it cannot be doubted that it lies within the cone, and also that it is formed largely by the layering of the media.

Due to the small size of the images, the above measurements lay no claim to great accuracy, rather they are intended to show that the images along the radius of a tangential section decrease in size and then increase again, whereby the variation in image distance is also demonstrated. The smallest images seen lay in the truncated surface of the cones and in vivo would thus lie within the cone.

4. *Calculation of the Dioptrics of the Lampyris Eye*

As we have seen, the crystalline cone together with the corneal facet functions as an astronomical telescope focused at infinity. The ray path in such a system follows from Text Fig. 14.

$$ab = ah_1 \cdot \tan\alpha = ah_2 \cdot \tan\beta$$

If we call the two focal lengths, ah_1 and ah_2 φ_1 and φ_2

$$\tan\alpha/\tan\beta = \varphi_1/\varphi_2 = \text{const.} \quad \ldots\ldots\ldots\ldots \quad \text{Eq. (1)}$$

which resembles Snell's law of refraction, and differs in that instead of the refractive index of the two media the two focal lengths are used, and instead of the sines of the angles the tangents.

If we now consider rays incident on the spherically curved eye, each being deflected according to Eq. (1) then the image can be constructed in an manner analogous to the construction of an image at a spherically curved boundary between two media. The equivalent of the radius, or normal, is the radius of the eye or the axis of the ommatidium; only in the latter case the incident and refracted rays lie on the same side of the normal, and we must ignore the thickness of the optically functional layer, i.e. the length of the crystalline cone and cornea.

The ray path shown in Text Fig. 16 has been constructed on these assumptions for angles up to 60°, when $\tan\beta/\tan\alpha = 1.5$. One sees that the rays converge onto a caustic conical surface not unlike the behaviour with spherical surfaces and the sine law. The solid lines show the ray path and the caustic curve for the tangent law, the dotted lines the same for Snell's law.

For the small angles in which we are interested

$$\tan\alpha/\tan\beta = \sin\alpha/\sin\beta$$

Using this substitution, I have performed the calculations totally in accord with the computation of the refraction at a spherical surface given by Helmholtz (1867). As expected, apart from the signs, completely analogous equations result; I have tried to make this obvious by retention of the standard symbols.

In Text Fig. 15 *bc* is the curvature of the eye, *ap* a straight line passing through the centre of curvature *a*, a ray leaving the point *p* reaches the eye at *c, ad* is the

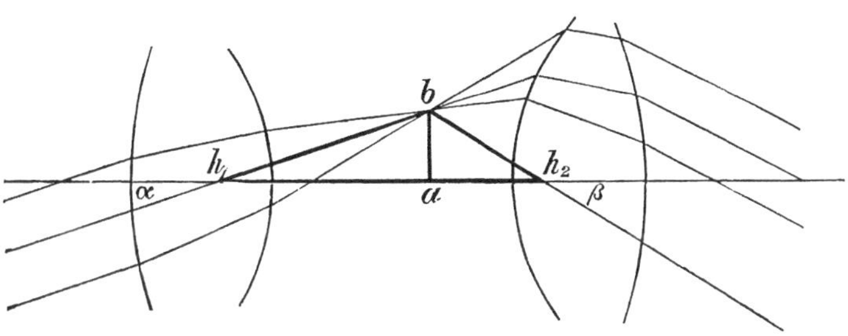

Text Fig. 14

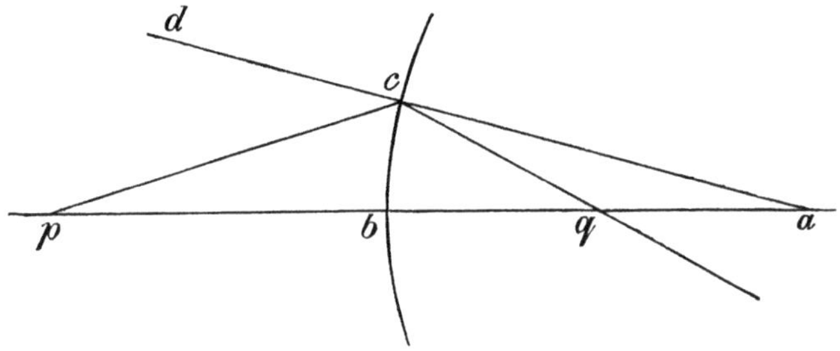

Text Fig. 15

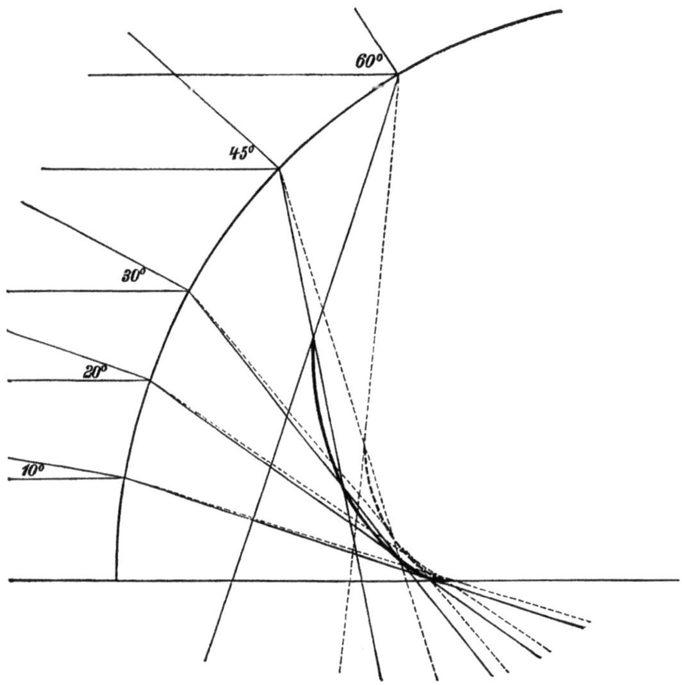

Text Fig. 16

normal for this ray, the refracted ray intersects *ab* at *q*. From Eq. (1) and using the above substitution:

$$\varphi_1 \sin pcd = \varphi_2 \sin acq$$

In the triangle *pca*

$$\sin pca/\sin cpa = ap/ac$$

and in triangle *acq*

$$\sin qca/\sin cqa = aq/ac$$

On dividing the first of these equations by the second and noticing that sin pca = sin pcd, one obtains

$$\sin pcd \cdot \sin cqa / \sin qca \cdot \sin pca = ap/aq$$

Now from Eq. (1)

$$\sin pcd / \sin qca = \varphi_1/\varphi_2$$

and in triangle pcq

$$\sin cqb / \sin cpa = cp/cq$$

so that, since sin cqa = sin cqb, one can also write the above equation as:

$$\varphi_2 \cdot cp/\varphi_1 \cdot cq = ap/aq \quad \ldots\ldots \qquad \text{Eq. (2)}$$

When the rays meet the eye near b then:

$$cp/cq = bp/bq$$

and Eq. (2) becomes

$$\varphi_2 \cdot bp/\varphi_1 \cdot bq = ap/aq \quad \ldots\ldots\ldots\ldots \qquad \text{Eq. (3)}$$

Let us now introduce alternative symbols for the distances of the two conjugated points q and p, first from the dioptric apparatus and secondly from the centre of curvature. These are all positive in the direction of the object.

$$ab = r \quad bp = f_1 \quad bq = -f_2 \quad ap = g_1 \quad aq = g_2$$

From Text Fig. 15 one sees:

$$f_1 + r = g_1 \ldots\ldots\ldots\ldots\ldots\ldots \qquad \text{Eq. (4)}$$

$$-f_2 = r - g_2$$

Then from Eq. (3)

$$\varphi_1/f_1 + \varphi_2/f_2 = (\varphi_1 + \varphi_2)/-r \quad or \ \ldots\ldots\ldots \qquad \text{Eq. (5)}$$

$$\varphi_1/g_2 + \varphi_2/g_1 = (\varphi_1 + \varphi_2)/r$$

If one designates the respective values of f_2 and g_2 with capital letters when f_1 and g_1 tend to infinity, then from Eq. (5)

$$F_2 = -r\varphi_2/(\varphi_1 + \varphi_2) \ldots\ldots\ldots \qquad \text{Eq. (6)}$$

$$G_2 = r \cdot \varphi_1/(\varphi_1 + \varphi_2)$$

and similarly when f_2 and g_2 tend towards infinity,

$$F_1 = -r\varphi_1/(\varphi_1 + \varphi_2) = -G_2 \ldots\ldots\ldots \qquad \text{Eq. (7)}$$

$$G_1 = r\varphi_2/(\varphi_1 + \varphi_2) = -F_2$$

from which it follows

$$F_1/F_2 = \varphi_1/\varphi_2 = G_2/G_1$$

From Eq. (6) it follows that when $\varphi_1 = \varphi_2$, i.e. when the astronomical telescope which could be substituted for the dioptrics consisted of two identical lenses, then the secondary focal point of the *Lampyris* eye would lie in the midpoint of the radius, i.e. just where the focal point of a convex mirror would lie.

The retinal image however, is a real image. From Eq. (7) one sees that the primary focal point of the eye also lies at the same point. From these two equations it further follows that

$$r = G_1 + G_2 = -(F_1 + F_2)$$

and further one can eliminate φ_1 and φ_2 by using the equations to transform Eq. (5) into the familiar form:

$$F_1/f_1 + F_2/f_2 = 1 \quad \text{Eq. (8)}$$
$$G_1/g_1 + G_2/g_2 = 1$$

from which it follows that

$$f_1 = F_1 \cdot f_2/(f_2 - F_2) \quad \text{Eq. (9)}$$
$$f_2 = F_2 \cdot f_1/(f_1 - F_1)$$

If, as in the case of the experiment described on p. 39, the rays come from the retinal side, then in Eq. (5)

φ_1 becomes φ_2
φ_2 becomes φ_1
f_1 becomes $-f_2$
$-f_2$ becomes f_1

so that this equation takes the form:

$$\varphi_1/f_1 + \varphi_2/f_2 = (\varphi_1 + \varphi_2)/r \quad \text{Eq. (10)}$$
$$\varphi_1/g_2 + \varphi_2/g_1 = (\varphi_1 + \varphi_2)/-r$$

i.e. the equation remains the same as with normal refraction [cf. Eq. (5)] except that the sign of the radius is altered. Thus the inverted image one sees with antidromic illumination lies in exactly the same place as the normal retinal image, when in both cases the object lies at a great distance.

With respect to the calculation of the image size, it follows from Text Fig. 17 that:

$$ps/rt = ap/ar = g_1/g_2$$

Calling:

$$ps = \beta_1 \quad \text{and} \quad rt = \beta_2$$

then

$$\beta_1/\beta_2 = g_1/g_2$$

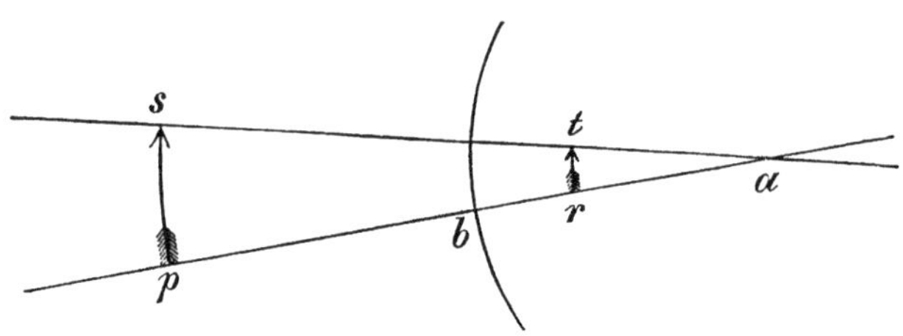

Text Fig. 17

and considering Eqs. (4–7):

$$\beta_1/\beta_2 = (g_1 - G_1)/G_2 = G_1/(g_2 - G_2)$$
$$\beta_1/\beta_2 = (F_1 - f_1)/F_1 = F_2/(F_2 - f_2)$$

If one considers the case in Text Fig. 17 when *rt* and *sp* move towards *b*, one sees that they meet at *b* with the same size and direction.

Thus it follows from the last equation that when $\beta_1 = \beta_2$ then $f_1 = 0$, i.e. both principal points lie in the refracting surface, or with respect to the insect eye: in every facet there is a common principal point and all these points are arranged on a spherical surface around the centre of curvature of the eye.

It is easy to demonstrate the general validity of these calculations, e.g. the predicted approach of the image towards the dioptric apparatus on approach of the object, and other details already mentioned in the above description of the optical behaviour of the eye. However, I did not succeed in obtaining a satisfactory agreement between the measured and predicted numerical values. Thus, for example, I measured the distance of the image from the crystalline cones for three object distances of 810, 12 and 1.2 mm. (I achieved the small distances by using images projected with Abbe's illumination system as objects.) From this one can calculate the ratio φ_1/φ_2 using Eq. (3). In each case I obtained a value between 1.1 and 1.9, which is in qualitative agreement with observations and the above description in that this also demonstrates that the inverted image which must lie in each cone does indeed lie in the proximal half. Nevertheless, the theoretical values differed significantly from the measured ones. The same was true when I tested the equations in which the measured focal lengths are introduced. The reason for this unsatisfactory agreement appears obvious. The experimental measurements were all made between the image and the relatively thick dioptric apparatus. As already explained, one cannot focus accurately at a particular level within the dioptrics – one needs only remember that one is dealing with images like those seen in Fig. 3. Only a small error here becomes very significant in the calculated result. Further, the calculations ignore the thickness of the dioptrics, and the experiments could only be performed in specimens preserved in alcohol. By contrast, my model of the insect eye gave result in very good agreement with the calculations.

Up to now I have assumed that the structure of the dioptrics of the firefly eye is that of a lens cylinder. In Chapter I, however, I mentioned that the step lens of Matthiesen can produce similar optical effects.

The question thus arises whether the function of the dioptrics is perhaps based on this principle. I believe the answer must be no, since to achieve the action of two convex lenses with such short focal length with this principle would require enormous, highly unrealistic differences in refractive index in the various layers, which would have to be immediately apparent under the microscope.

The image formation in *Lampyris* can also be formulated in a different way as follows. Each cone has proven to be an astronomical telescope, and as such projects an upright image on the retina. If the telescope is focused at infinity, i.e. the rays originating from a common point emerge parallel, then the cross-section of such a ray cylinder will be small compared to dimensions of the image which will thus still be rather sharp. These erect images overlap so that, according to Figs. 2 and 3, for each point on the object approximately 30 retinal images are superimposed. For a second point on the object another 30 facets, some of which – depending on the separation from the first point – may be the same, compose the definitive retinal image. Ignoring the changes in optical parameters at the eye margin and other small variations, these 30 images have the same size and shape and differ only in their position. Facets lying more to the right still contain parts of the object lying to the right, and those on the left more from the left side of the object. In order to characterize this type of image, and to distinguish it from the single images of the individual facets which compose it, I have thus called it a "superposition image".

C. Catoptric Function of the Cone

In the accompanying plates of the ommatidial structure in various arthropods, one notices that the distal surface of the crystalline cone in insects is almost always markedly smaller than the proximal surface of the corneal facet. It is thus possible that, after passing through the cornea, rays may travel alongside the cone rather than entering it. In fact there is no doubt that such rays must exist. They are those which meet the eye with too great an angle of incidence.

Normally there is no doubt that these rays will be absorbed by the iris pigment lying between the cones. However, it seems questionable if this is always the case, indeed under certain positions of this pigment (light-adapted position – see Chap. III) it is not obvious why these rays should not first penetrate the neighbouring crystalline cones (cf. Fig. 16B or Fig. 52 in a crustacean).

In this respect I wish to describe a mechanism which probably plays an important role in the screening of aberrant rays in insect eyes. It also gives an explanation for the almost universal occurrence of one of the most important components of the dioptrics, namely the crystalline cone. I refer here to the possible reflections of rays in the cone.

Every ray undergoes reflection at the boundary between two optical media. The reflection is usually partial, but when a ray travelling in an optically denser medium meets an optically less dense medium it may be total. Reflection is total when the angle of incidence exceeds a certain value determined by the refractive indices of the two media. Whether the reflection is total or partial, the angle of reflection always equals the angle of incidence.

Crystalline cones derive their name from their cone-shaped form, and cones behave quite characteristically with respect to the reflection of internal rays. Text Fig. 18 shows an ideal cone, *a* is a ray incident with a modest inclination. At *b* it meets the first boundary with an optically less dense medium. The angle of incidence here is so large that reflection will probably be total. The same may be true for point *c*. Simply by equating the angles of reflection and incidence, the path of the reflected component of the ray can be constructed as *d, e, f, g, h, i*. Thus as long as it is reflected the ray will be forced back out of the cone. However, a larger part of the ray will no longer be reflected at *d, e, f, g* but will penetrate

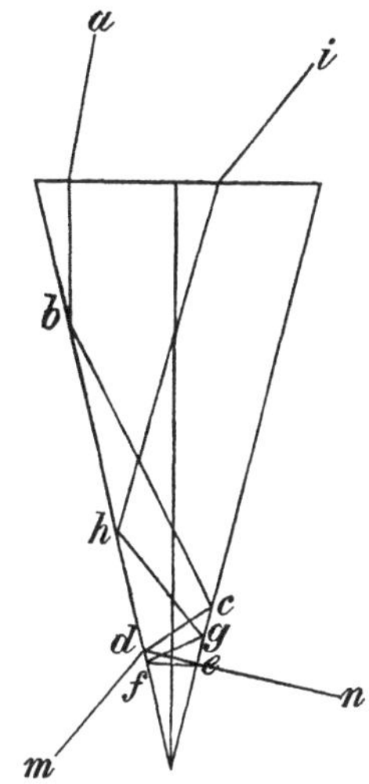

Text Fig. 18

the boundary between the two media instead. These ray components leaving the cone (*dm, en* etc) now have a very significant inclination to the cone's axis and there is now no ray component travelling even approximately in the same direction as the cone axis.

If we apply these considerations from a geometrical cone to the cone of the compound eye, which does not have a geometric apex but a curved surface, and which is also not homogeneous but layered as a lens cylinder, then it is clear that rays can only emerge at the apical surface as already discussed at length. In the geometrical cone this only true for the infinitely thin axial ray.

Although the cone of the compound eye differs from a geometrical cone in having a lens cylinder construction and an apical surface through which rays can emerge, it is by no means excluded that certain rays in the crystalline cone also experience the same fate as those in the geometrical cone. Light that is reflected back along the path *a, b, c, d, e, f, g, h, i* will be negligible, but there will certainly be rays which, like *dm* and *en*, emerge from the crystalline cone at such an angle that they must either be absorbed by the iris pigment (see figures mentioned above) or penetrate a neighbouring cone. In the latter case, they already enter the second cone with a very large angle to the cone axis, and thus immediately suffer the same fate again. In this way they may perhaps pass through several cones, but they will finally be ejected out of the whole optical system, as with *hi* in the geometrical cone. The refraction experienced by the rays upon crossing the cone boundaries also aids in this process. (Indeed, diagramatically a radial section through the cones would look like an array of prisms with their refracting edges facing backwards, which would, of course, deflect any ray passing through them forwards).

Naturally these considerations do not apply to those rays which we dealt with in the sections on dioptrics. These undergo no reflection. However, rays incident with a sufficiently large angle to the axis will intersect the lateral boundaries of the cone where they will be reflected or refracted accordingly.

I believe a phenomenon which had long puzzled me can also be explained by the foregoing. Many insects, e.g. *Hydrophilus*, and especially crustaceans, show a characteristic diffuse eyeshine which appears to have a very superficial origin. This is particularly obvious when the animal is brought into direct sunlight. The rays reflected back out of the eye as described are then simply more intense. This irridescence is seen best in the large crayfish and lobsters, where the intensity distribution in the reflections delineates the shape of a right-angled cross extending over a large part of the eye. I am probably not wrong in correlating this with the right-angled facets and square crystalline cones in these animals.

These considerations may also explain the significance of the strongly reflecting granular material lying distally from the iris pigment in certain crustaceans and insects. It moves with the iris pigment following illumination and appears identical to the material in the tapetum of these crustaceans. In transmitted light it appears almost as black as the iris pigment, and appears to be part of it, but in dark field one recognizes it as an opaque strongly reflecting substance. It is present in *Nica edulis, Sicyonia sculpta, Crangon, Palaemon* and also in *Palinurus*, though not so clearly. I call this material the iris tapetum by analogy with the coloured pigment cells which are commonly seen in distal sections of ommatidia in insects.

The significance of this iris tapetum seems reasonably clear. Since it lies between the cones, or their mantles, it will reflect those light rays which reach the spaces between the cones as just discussed. It thus reinforces the dioptric and catoptric arrangement just described in driving out light which has penetrated the ommatidium too obliquely. In both insects and crustaceans it gives rise to the, often, spectacular eye colours, and reflections from it explain the beautiful metallic eye shine which, particularly in insects, is no rarity.[20]

Previously, the iris tapetum appears only to have been noticed by Leydig (1855) in the crayfish. He writes:

"About halfway between the end of the crystalline cone and the tip of the spindle-shaped swelling one also encounters, apart from the black pigment, a pigment which appears white in incident light and is also composed of small granules."

As I will show shortly, he was thus looking at an eye with iris tapetum and iris pigment in the light-adapted state.

Chapter III The Iris Pigment and its Function[21]

It is certainly significant that the eye of *Phronima* (Chap. VIII), which projects an image according to catoptric principles, contains no pigment, whereas I know of no compound eye with a dioptric function which does not contain two well-defined pigment layers. By pigment I refer to the granular black pigment, similar to that found in vertebrates, and disregard the coloured diffuse pigment and abnormal albino conditions which also occur amongst insects and crustaceans. The most distal pigment layer lies within or near the dioptric apparatus, and I call it the iris pigment on functional grounds. The second layer lies on or between the elements of the retina and even behind it. I call this retinal pigment and it will be considered later.

Figure 1 of the *Lampyris* eye shows the iris pigment filling the spaces between the crystalline cones; only the tips of the cones remain free and project into the transparent material between the retina and the dioptrics.[22] The figure is from a specimen killed in the dark. The iris pigment occupies a different position in animals exposed to the sun. After it has become motionless, the animal is left a further one or two hours in the sun before being brought into a different illumination to ensure that the internal organs have lost their ability to react.

In such a light-adapted *Lampyris* eye, the majority of the iris pigment lies proximal to the cone tips and only sparse traces remain between the crystalline cones. The pigment has thus migrated proximally, approximately the length of the crystalline cones. However, it does not envelop the apical surfaces of the cones – in this case no light would reach the retina – but the extension of the cone axis and the immediate surroundings remain free from pigment, as is readily seen in oblique and cross-sections. Thus the pigment moves over the cones more or less as a cylindrical sheath, and together the sheaths of the various cones form a coherent pigment mass.

With a knowledge of the ray path in the *Lampyris* eye (see Chap. II) the functional significance of this pigment migration is obvious. Text Fig. 19 is a diagram of the eye showing the dark-adapted state below and the light-adapted state above. Rays of light incident upon the eye from *a* are refracted through the individual facets and converge at *b* to form the retinal image. In the light-adapted state some of these rays must be absorbed by the pigment and thus never reach *b*. Further, with the gradual transition of the pigment from the dark- to the light-adapted state, more and more rays from the ray cone (apex *b*) must be intercepted. The base of the cone thus becomes progressively smaller and the relative brightness of the image decreases proportionally. The mammalian iris reduces the brightness of the retinal image similarly, by progressive contraction

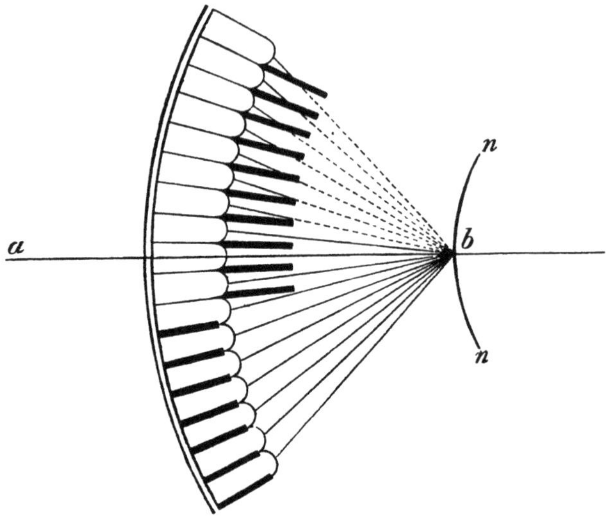

Text Fig. 19

of the pupil. Again starting at the outside of the cone base the pupil intercepts ray after ray, reducing the brightness of the point *b* in proportion to the cone base. Exactly because of this functional analogy I propose the name "iris pigment" for this pigment layer.

Just as in the vertebrate eye, the compound eye can apparently adapt to a wide range of intensities by means of its iris pigment. Shortly I will cite examples which suggest that the ray cone can be reduced to the rays from a single facet, whilst the pupil of the vertebrate eye rarely achieves an analogous reduction. Only amongst sharks does the slit-shaped pupil appear to close completely, leaving only a hole at each end of the slit. Even the smallest pupil of the cat is less effective than the iris pigment of numerous arthropods.[23]

Such an effect of the iris pigment is, of course, only possible in superposition eyes. As in *Lampyris*, there must be a considerable zone between the dioptrics and the retina through which rays can travel and into which the pigment can migrate. In apposition eyes, e.g. *Limulus,* such a pigment migration would be pointless. Conversely, in doubtful cases one can take the presence of pigment migration under illumination as an indication that one is dealing with a superposition rather than an apposition eye.

I have checked for the presence of such pigment migration in numerous arthropods, and very significantly I have only found it amongst nocturnal animals, i.e. animals which use their eyes by both night and day. Moths are not blind by day, and most crustaceans are nocturnal, but still see very well by day. These all show pigment migration. Flies, butterflies, dragonflies etc. on the other hand, are completely helpless at night, show no effect of light on the iris pigment and, as will be described later, have apposition eyes. I also studied dragonfly larvae and found no effect of light on the iris pigment.[24]

Before presenting my findings on the role of the iris pigment in a wide range of animals I should point out that certain discrepancies in the literature can be explained by the state of adaptation. For example Leydig (1864a) shows a diagram of the hawkmoth eye, with the iris pigment changing gradually from a dark-adapted state on one side to an intermediate state on the other. The animal must have died with the eye receiving a retinal image which was dark on one side and bright on the other. Thus I was able to demonstrate that the action of light on the iris pigment is local and could set one half of the eye in the light-adapted state and the other in the dark-adapted state by making half of the visual field bright and the other half dark. One can thus makc "photographs" on the iris pigment, just as Kühne made "optograms" with vertebrate retinae. I will discuss this further under the topic of eyeshine (Chap. IX) as these properties can be studied on live animals with the aid of eyeshine much more easily than through histology.

I will also discuss then, the time course of the pigment migration; here it is only emphasized that, for the anatomical studies, I kept the animals for 1–2 hours or overnight in a totally dark room where they were killed by immersion in alcohol, and other specimens were treated similarly after being placed in sunlight. Without going into details, it seems that both dark- and light-adapted states are more pronounced when the animals are not exposed too long to dark or light. In particular, several days in the dark can result in an intermediate pigment position. This would agree with the recent findings of Fick (1889) on the retinal pigment in frogs. The health of the animals should also considered and it is my experience that weak animals on the verge of dying no longer show pigment migration.[25]

A. Insects

1. Beetles

As well as in the firefly, pigment migration can also be demonstrated in *Cantharis fusca* (*Telephorus*) (Fig. 9). The eye of this beetle also projects an unmistakable superposition image, and the position of the pigment in the dark-adapted eye is very similar to that in *Lampyris*. After exposure to sunlight, the pigment is shifted proximally, with virtually none left between the cones, whilst in the dark-adapted eye the space between the cones is completely filled with pigment.

I investigated one dark- and two light-adapted eyes of the rose chafer (*Cetonia*) but found no difference in the position of the iris pigment although the structure of the eye would otherwise lead one to expect it. Thus the pigment ends with an abrupt boundary at about the level of the cone tips, filling the space between the cones. Since, at that time, I had not recognized the difference between diurnal and nocturnal eyes, I could only attribute the lack of pigment to interspecific variation, particularly as, in other species, I had only seen such sharp pigment boundaries in dark-adapted eyes. I thus also sectioned the eye of a smaller close relative of the rose chafer, *Tropinota hirtella* (Fig. 23). The eye is very similar, and although I looked at three light- and three dark-adapted eyes, I again found no trace of pigment migration. The rose chafer and their smaller relatives are both extremely diurnal, flying in the brightest sunshine and spending the night concealed in flowers, etc., apparently asleep. This is probably correlated with the lack of pigment migration. It is only remarkable that the eye structure, with an unpigmented space between the retina and the cones, is suggestive of a superposition image. Of course, the question of the location of the sensitive layer of the retina remains: a superposition image can only be considered if this lies in the proximal elongated swellings of the retinulae, but not if it corresponds to the distal button-shaped ends. Other very close relatives of these beetles, e.g. the June and cockchafer, are nocturnal. I would thus conclude that *Cetonia* and *Tropinota* have eyes which have evolved from nocturnal eyes like those of the related cockchafer. The diurnal habit is secondarily acquired, the eye retained its superposition image but lost the ability to vary the brightness of the retinal image through the iris pigment.

The evolution of the cetonians can thus be inferred from their eyes. Their diurnal nature is also indicated, e.g., in Taschenberg's introduction to the chapter on these beetles in Brehm's "Animal life":

"The last ... genus (of the Lamellicornia) comprises the honey-loving Cetonidae... animals which do not shun the light and wait for the cover of night to come out of hiding, but which seek out the scented flowers and flowering bushes in broad daylight, feeding on the pollen together with the spectacular butterflies, the lively flies and the ever busy bees".

I also succeeded in demonstrating pigment migration in two diving beetles and one water beetle. These were the first experiments which I made to check my supposition of pigment migration (Exner 1889).

In *Dytiscus marginalis* the iris pigment fills the space between the cones in the dark-adapted eye, leaving the tips free so that the rays can emerge with a considerable inclination to the axis (Fig. 7). In the light-adapted eye (Fig. 6) there is still copious pigment between the cones, but a large proportion is proximally shifted and covers the retinulae for approximately the length of one crystalline cone. In this situation only rays more or less parallel to the optical axis of the cone can emerge and travel further without being absorbed by the pigment sheathes. I found almost exactly the same behaviour in *Colymbetes fuscus* (Fig. 16).

Although somewhat different in outward appearance, the photomechanical behaviour of the iris pigment in *Hydrophilus piceus* has the same optical effect. I have sectioned numerous light- and dark-adapted eyes of this animal, as it shows a peculiarity which renders the mechanical process of pigment migration easier to understand. In the dark-adapted eye (Fig. 13) the pigment is so densely packed between the cones, along their entire length, that very thin sections are required to see the cones at all. However, horizontal sections reveal that each cone tip corresponds to a round opening in the pigment through which a superposition image can be projected onto the rather distant retina. In most species, the space between the cones and the retina is filled by the retinulae or their narrow proximal extensions, which join onto the cone tips. But in the *Hydrophilus* eye there is no trace of this radial pattern; indeed, I could find no connection between the retina and the dioptric apparatus. Instead, the space is filled with numerous fibres running parallel to the surface or rising obliquely. They have a succulent appearance and are possibly muscular in nature.[26]

At the edge of the eye these fibres tend to radiate out towards the pigment layer in a fan-shaped array; I have also found places where they ascend radially from the retina directly towards the iris pigment. It might be considered that all the fibres originally had this orientation and that the layering parallel and oblique to the cornea is an artefact of preparation, but I consider this excluded – the length of the fibres alone is too great for such an interpretation. Distally the fibres merge into the pigment, with small cone-shaped processes forming direct contact with the pigment layer. I could only recognize the proximal end of the fibres at those places where they ascend more or less directly to the pigment. Here they appear to make intimate contact with the retina.

In the light-adapted eye (Fig. 14), only a small amount of pigment remains between the cones, appearing to lie in branched processes of the pigment cells, and one can see the cones easily, even in thick sections. The pigment is now concentrated at the level of the cone tips and reaches a considerable distance proximally. Since the dimensions of not only the whole eye, but also the retina, are unchanged, the space between the pigment layer and the retina, i.e. the space occupied by the fibres, must have diminished. The fibrous mass is compressed together and the cone-shaped, pigmented ends of the fibres have shifted proximally with the pigment layer. One could assume that the pigment cells, perhaps using the cones for support, have expanded proximally, compressing the fibres with which they are fused; but it seems more likely that the fibres, which originate more or less directly from the retina, have contracted, thus pulling the pigment cells with them. The latter hypothesis is strongly supported by observations in moths discussed below. If true, these fibres must be considered as organic muscle fibres. Indeed, Leydig has described and illustrated such fibres in moths.

I thus find it probable that the action of light results in the iris pigment being pulled backwards by muscle fibres. Even so, there are animals in which one would have great difficulties in establishing the fibres anatomically despite the presence of pigment migration. This is particularly true for crustaceans, in which, despite the most striking photomechanical effects, I could find no trace of muscle fibres.

2. *Lepidopterans*

As expected, pigment migration could only be demonstrated in moths; however, here it is exceedingly pronounced. In the dark-adapted eye, each presumed muscle fibre appears to be attached to an elongated pigment cell, lying between the cones and barely projecting beyond their tips (Fig. 29). On light adaptation, the pigment cell is further elongated and shifted a long way proximally. The distal end still lies between the cones but the main mass lies between the retina and the dioptric apparatus (Fig. 28).

The following experiment I performed twice on the lappet *Lasiocampa (Gastropacha) quercifolia*. The animal was kept in the dark, then one half of the head (with the eye) was removed and placed in alcohol. The animal, which appeared hardly damaged, was then placed in the sun for about half an hour before being killed in alcohol. In section the eyes now showed very extensive pigment migration (Fig. 28 from the second eye, Fig. 29 from the dark-adapted eye). I performed a similar experiment with the same result on one other moth (probably *Leucoma*).

I have further demonstrated pigment migration in a whole range of smaller moths (*Spilosoma* or *Porthesia* and their brown relatives) in the usual way by taking sections of eyes from two animals, one killed in the dark and the other in the sun (Figs. 30, 31). In every case one sees fibres merging into the pigment cells. These can be interpreted as muscle fibres and are analogous to the muscle fibres which are known to distend the chromatophores in cephalopods, thus causing colour changes in the skin. These fibres were first stimulated electrically by Brücke (1852), and more recently have been investigated by Klemensiewicz (1878) in more detail.

À propos the local effect of light already mentioned, if the conditions of illumination are not controlled, one occasionally finds one part of the eye in the dark-adapted and another in the light-adapted state. I saw this for example in the red underwing (*Catocala nupta*, Fig. 15).

B. Crustaceans

The action of light on the iris pigment can be equally clearly demonstrated in a number of crustaceans. A sojourn at the zoological station in Naples enabled me to study a considerable number of these. In the large majority of animals I found unmistakable photomechanical effects, but in a few I failed to find any. This difference is almost undoubtedly correlated with the animal's way of life, but I cannot prove this since in the literature I could find no information as to which species were nocturnal and which diurnal. The situation is further complicated since the intensity for which the eye of a marine crustacean must be adapted varies with the depth at which the animal lives. It is thus to be expected that the eye can adapt to a variety of intensities, even in an exclusively diurnal species, if the animal has the ability to live and chase its prey at various depths.

I have observed migration of the iris pigment under the action of light in the following: *Palaemon* (Figs. 51, 52). In the dark the pigment lies entirely between the crystalline cones. By cones I refer to the strongly refractive parts of the dioptric apparatus, which, from their location and presumably also function, correspond to the cones in insects. Schultze (1868) considers the entire, cone-shaped segment between the cornea and the retina as the crystalline cone and distinguishes several sections within it.

In *Palaemon* the iris pigment has very sharp distal and proximal borders and adjoins the cones tightly in a thin layer. Even with the naked eye, one can recognize the very considerable change in position of the pigment in sections of the light-adapted eye. The pigment now forms a dark zone lying far behind the cones, and has in fact moved into the vicinity of the retina. The cones are completely free of pigment. Both proximal and distal borders of the pigment are also sharp in this new position. In cross- or oblique-sections (with respect to the ommatidial axis) the pigment layer is seen to be pierced like a honeycomb.

Characteristic filaments pass through the holes in groups of four. These are extensions of the crystalline cones which, themselves, comprise four prismatic sections. I also sectioned an eye from a specimen of *Palaemon* which was kept and killed in the back of a room not illuminated by the sun. As was to be expected the iris pigment in this twilight-adapted eye was in an intermediate position.

Although this type of pigment migration differs qualitatively and quantitatively from that seen in insects, it is quite obvious that the physiological function is essentially the same. As the pigment moves closer to the retina, the cone of light reaching the retina from the tip of each crystalline cone becomes narrower, and in the most extreme case perhaps only one or a few retinulae will be illuminated by each cone. Should only one be illuminated, the situation would correspond to Müller's mosaic vision, except that the cone concentrates the rays from a point on its axis to an approximately parallel beam, so that the slender retinal element is more intensely illuminated.

In the spiny lobster, *Palinurus* (Fig. 50), I could detect no obvious difference in the position of the iris pigment of an animal killed in diffuse daylight and one kept in the dark. However, the first animal had not been in direct sunlight and the second was killed under weak illumination lasting at least several seconds, this being necessary because of the strength of these large animals. I hardly doubt that under more favourable conditions differences in the iris pigment could be revealed.

Nica edulis gave very striking results, but as the eye is so similar to that of *Palaemon* and *Sicyonia* it is not illustrated. In the dark, the iris pigment lies between the cones, leaving their proximal ends free. After illumination, all the pigment has moved behind the proximal ends of the cones, thus resembling the situation in *Palaemon* except that single pigment strands project even further proximally.

The eye of *Sicyonia* is extremely instructive (Figs. 53, 54). Illumination results in the pigment migrating halfway down the eye, so that the layers of the eye appear to have changed completely when viewed at low magnification (Fig. 39: *D* dark-adapted, *L* light-adapted, *A* light-adapted eye seen in incident light). In the last section (*A*), only the tapetal layers are visible against a dark background and one can see that the iris tapetum has migrated along with the iris pigment, as is the case in other crustaceans.

The local fresh-water crayfish (*Astacus fluviatilis*) also only shows a pronounced photomechanical effect on the iris pigment when it is killed in good condition, but here not all the pigment moves on illumination. Sparse amounts remain between the cones, whilst the majority migrates in long sheaths along the cones and up to the retina.

Amongst the macrurans only the bulldozer shrimp, *Scyllarus* (Figs. 25, 26) and *Peneus* failed to show migration of iris pigment. The specimens of *Peneus* which I studied possessed virtually no iris pigment anyway, and in *Scyllarus* pigment migration could probably be detected if one undertook new experiments. Unfortunately, it was not possible to check my results as I preserved all the eyes of marine crustaceans in Naples and first examined them under the microscope in Vienna. My results may thus be partially invalidated through particular sensitivity or insensitivity of the animals to light or other relevant factors.

The hermit crab, *Pagurus*, showed very obvious pigment migration. In the dark the pigment lies exclusively between the cones, leaving most of their proximal halves and the tips projecting freely backwards. On illumination, it moves along the cones towards the retina but without completely clearing the space between the cones. This is in keeping with its position between the macrurans and the brachyurans. Thus in the brachyurans the pigment does not migrate en masse, as is the case in most macrurans, but, rather, there is a proximally directed "pigment expansion".

A glance at the illustrations from light- and dark-adapted crab eyes shows what I mean. Figures 55 and 56 show two eyes from *Dromia*, the former kept in the dark and the latter in diffuse daylight. Even under this moderate illumination, the pigment has moved a large distance proximally, but remnants are left behind. Further, and this is very characteristic in these crabs, the iris pigment merges proximally with the retinal pigment (which has migrated distally) to form a continuous, though thin, pigment sheath around the retinula and its prolongation.

In *Maja* (Figs. 41, 42) the pigment migration is similar though not so extensive. The nuclei in the distal parts of the retina are surrounded by pigment in the light and almost free of pigment in the dark.

Galathea (Figs. 45, 46) behaves very similarly to *Dromia* and also shows an intermingling of retinal and iris pigment.

Such an intermingling also occurs in *Pisa* (Figs. 48, 49), but this species is distinct from the other crabs in that (as in the macrurans) on illumination all the iris pigment leaves the space between the cones. This pigment migration is all the more conspicuous since the distal swelling of the retinula also moves a considerable distance proximally. One then sees a long connecting fibre between the cone tip and the retinula which is not visible in the dark-adapted eye.

The pigment migration in *Portunus* is more distinctive (Figs. 37, 38). The pigment moves as a rather compact mass a short distance along the retinula and surrounds the distal swelling.

Carcinus maenas also showed unquestionable photomechanical effects, but in *Squilla mantis* I could not detect any movement of the pigment.

Chapter IV The Retinal Image of Various Insects and Crustaceans

Despite their enormous diversity, almost all the arthropod eyes which I have studied can be traced back to one of two types, exemplified by the compound eyes of *Limulus* and *Lampyris*. In the mixed forms discussed below, which could be considered as special forms, I see rather a confirmation of the concept of two basic types.

The most fundamental and anatomically easily established difference between the two types is in the position of the retina, at least insofar as: (1) a superposition image is only possible in eyes having a broad zone of transparent tissue between the dioptric apparatus and the sensitive layers of the retina so that a retinal element may be reached by rays from several cone tips, and (2) an eye with retinal elements in the direct vicinity of the cone tips can project an apposition image.

In many animals, the retinal elements extend so far proximally from the cones that their distal parts could receive an apposition image whilst the proximal parts could quite possibly receive a superposition image. In fact, as discussed below, this is probably partially true. For the moment, though, it suffices to say that a retinula adjoining the cones is indicative of an apposition image and a remote retina suggests a superposition image.

The two eye types can also be distinguished by their optical properties without extensive investigation. Normally when one slices off a fresh or a fixed eye, the tissue behind the dioptrics, particularly the pigment, obscures a view of the retinal image. However, usually at least a pair of cone tips are exposed so that one can see the rays leaving them. If one takes two bright points as the object then: (1) in a superposition eye, the two ray bundles diverge proximally to the cones, cross over in the region of the cone tips and can be seen to diverge again distally to the cone tips; proximally to the cones, the right hand ray bundle disappears if one screens the right hand point source; (2) in an apposition eye the ray bundles diverge only slightly if at all, and the left hand ray bundle disappears on screening the right hand point. If this method fails, one can also use a single point source. A bright point corresponding to each cone can then be seen at about the level of the dioptric apparatus. By focusing towards the retina one recognizes the advancing ray bundle either as strongly converging (in a superposition eye) or as hardly converging at all (apposition). The origin of this behaviour can be understood from the ray paths described in Chapter II.

Naturally it is best when there is no tissue obscuring the image, and one can determine the eye type directly by checking the presence or absence of a superposition image under the microscope. For this purpose the animal is kept for one or two hours in the dark and then the eye is sliced off at such a level that the retinal pigment remains behind. When viewed in a hanging drop under the microscope, one then sees the bright apical surfaces of the individual cones. On focusing up, the blur circles of these light points flow together into a superposition image if such is present. Admittedly this tends not to be very clear since detached pigment, irregularities in the loose cones and probably also cloudiness in the tissue disturbs the image. However, assuming it exists at all, it can still be frequently recognized especially on moving a dark object (pencil) to and fro between the eye and microscope mirror. In the case of an apposition image one sees how the bright spots representing the individual facets disappear row by row in the sense of an erect image.

A. Superposition Images

1. Beetles

I deal first with those beetles which, like *Lampyris*, have the crystalline cone fused with the cornea. The demonstration of a superposition image is simple with these, as the intact dioptric apparatus can be brushed clean.

Such a fusion is already known in *Telephorus* and *Elater noctilucus*. My brother, Prof. Franz Exner, provided me with a specimen of an elaterid beetle (*Agrypnus moertus*) which he caught in India and preserved in alcohol. This also had the fused dioptric apparatus, and its eye, although now kept over a year in alcohol, projects the most perfect superposition image I have ever seen. It is qualitatively similar to the image in *Lampyris*, but many more cones are used in the production of a single image point, and the image is thus brighter. Like the beetle itself, the eye is large and I could see the erect retinal image with the naked eye and also the virtual inverted image located in the same position when the eye is illuminated in the reverse direction. From what I have seen of the eye of the American species, *Elater noctilucus*, I do not doubt that this beetle will have similarly beautiful retinal images.

The firefly *Luciola italica*, which I caught in Amalfi, also showed an image which was perhaps slightly clearer than in *Lampyris*, but which arises in exactly the same way.

In *Cantharis fusca* I again confirmed the report of a fused dioptric apparatus. I observed a fine superposition image, similar to that in *Lampyris* except that fewer cones are involved in the projection of an image point. I also established the presence of the inverted image with the ray path reversed; this is the simplest proof that the same optical mechanism is at work, although I did not neglect the experiment with the point object.

I found the same behaviour in *Cantharis rustica*. I also found the same anatomical situation in a related beetle, *Rhagonycha melanura*; I do not know if the fused cone and cornea were already known in this case. The optical study gave the same result. In this very small animal with a strongly curved cornea, the number of facets contributing to one image point is naturally rather small.

It is much more difficult to see the superposition image in insects without fused cones. The problem is to slice the eye so that the cones remain in their natural position and no disturbing tissue lies between the cone tips and the microscope. This latter problem can be overcome by exploiting the migration of the iris pigment .

Accordingly I proceeded as follows, e.g. with *Hydrophilus piceus*: the animals were brought out of their illuminated cage into a dark room (in water), after one or two hours the heads were cut off, and, protected from direct contact, placed in a freezing mixture. When I was sure that the eye was frozen, I brought it into the light and, using a cooled blade, made a shallow slice (so as not to include any part of the retina) which was then brought under the microscope hanging on a drop of beetle's blood as previously described. After thawing out, I could confirm the presence of the superposition image and also the virtual inverted image. Neither was particularly clear but that was only to be expected under the circumstances.

The eye of *Hydrophilus* differs quite significantly from the beetles mentioned so far, but shows many similarities with other insects (see also Exner 1875). The main difference is in the thickness of the cornea and its optical power. The cornea is not drawn in the figures of *Hydrophilus* (Figs. 13, 14), but the facets are approximately the size shown in Fig. 23. The corneal facet is a lens cylinder, spherically curved at both ends. Since the corneal surface is smooth the radius of curvature of the facet surface is identical with that of the corneal surface as a whole. It varies slightly in different sections and both ophthalmometrical and anatomical measurements give a value of about 1.4 mm. The radius of curvature of the hind surface was 0.013 mm (measured using a projection of the microscope image). With air at the front and water at the back surface, the corneal facet projects an inverted image 0.037 mm behind the back surface. Thus the rays entering the crystalline cone are strongly convergent. The cone is also built on the principle of a lens cylinder but it appears to have less refractive power than the cornea.

I confirmed this by observing that the cone still projects clear images after removal of its tip, and by showing (with the microrefractometer) that it deflects light entering perpendicularly to its axis as if it had a refractive index of 1.559 (assuming an homogeneous body, and averaged from several cones). If we recall that the exceedingly dense chitin of the *Limulus* eye has a refractive index of only 1.535, then it is likely that the strong refraction comes from the layering of a lens cylinder. Furthermore, truncated pieces of cones without apical surfaces still show clear ray convergence with the microrefractometer, and indeed project clear images, so that there can be no doubt about their being lens cylinders.

Thus here, also, the corneal facet and crystalline cone together form an astronomical telescope, the refractive power arising partly from the curved surfaces and partly from the layered construction. The most powerful component is the corneal facet which functions as a lens with very short focal length. Under natural conditions the cone is hard up against the back surface of the cornea and since it has a concave front surface the image must be shifted proximally. Thus, without any doubt, the image lies in the cone as in *Lampyris*, and the location of the image corresponds to the focal point in a lens cylinder twice as long as its focal length (see Chap. I). In this beetle the cone tip probably takes over the main function of the second half of the lens cylinder as a curved surface. Since it has a spherical refractive boundary with a very small radius, it must be attributed a strong action, particularly as the lens cylinder action of the cone is not very significant.

The proximal part of the corneal facet is considerably broader than the most distal part of the cone so that rays can pass through the cornea without penetrating the cone at all. These rays are either absorbed by the iris pigment or reflected back out of the eye, thus giving it its characteristic shimmer (see also Chap. IIIC).

Recognizable superposition images can be seen in the rose chafers, *Cetonia affinis, C.metallica*, and the related *Tropinota hirtella*. The quality of the images was good when I succeeded in cutting off the eye without disturbing the cones and at such a depth that the retinal pigment was excluded. In some specimens of *Tropinota* the image was almost as clear as in *Lampyris*. The fibre mass stretching between the cones and the retina must, of course, reduce the quality of the image since this tissue loses the transparency which it undoubtedly has in vivo. Nevertheless, I succeeded in seeing all the images illustrated for *Lampyris* (Figs. 2–5). One can see the rays converging into an image point which, although somewhat vague, even shows the same interference patterns as in the firefly. After *Lampyris*, the rose chafers were the first insects in which I found an erect retinal image. This is probably due to the fact that the iris pigment ends with a sharp boundary approximately at the level of the cone tips (see Chap. III), which thus project into the completely pigment free "vitreous". In other insects this only tends to be the case after dark-adaptation. This obviously aids the detection of the retinal image under the microscope.

The rose chafers and their relatives have very thick corneal facets which accordingly function as strong lens cylinders. If one brushes out a *Cetonia* eye, one again naturally sees an inverted image of an external object in every facet. This lies very close to the back corneal surface, perhaps slightly distally. With two point sources it could be established that the axes of the two very narrow ray cones running proximally from the image point hardly diverge. Thus the corneal facet corresponds to a lens cylinder of approximately the same length as its focal length (even when embedded in damar lacquer the image in the corneal facet lies very close to the hind surface). The remaining optical action presumably derives almost exclusively from the spherically curved apical surface of the very small, almost rudimentary, crystalline cone. The small size of these cones also prevents the rose chafer from experiencing excessive retinal excitation, even in harsh sunlight, despite its having a superposition image and apparent lack of iris pigment migration. (The significance of the relationship between the size of the facet and the cone has already been discussed on p. 40) *Oryctes rhinoceros* (Linn.) also has a very similar eye structure and optical behaviour to *Cetonia* and *Tropinota*.

2. *Lepidopterans*

Superposition images are only to be expected amongst the nocturnal species.[27] I did not go to the extent of making frozen preparations here, since superposition images could be demonstrated in preserved specimens kept in the dark. I proceeded as usual and cleared the opaque tissue of the clear zone somewhat with glycerine.

Admittedly, the image quality was not good. I had to be content with a vague light point, the image of a candle flame, moving with the candle in the sense of an erect image, or to check that, with one point source, the rays leaving one crystalline cone converge proximally, or that with two points, the two rays leaving one crystalline cone diverge proximally in the sense of a superposition image.

I confirmed this in *Sphinx convolvuli* and in four small moths: *Porthesia, Ocneria rubea, Leucoma salicis* and *Cnophria quadra*.

From this it seems justified to make the general statement that those moths with eucone eye where at least the broader parts of the retinulae form a well-developed retina some distance from the dioptric apparatus (this includes for example the day-flying hawk moth *Makroglossa*) have superposition eyes. Only then can the highly developed motility of the iris pigment in these animals be understood. As already emphasized, I must thus consider the demonstration of this characteristic pigment migration as reasonable grounds for the assumption of a superposition image. The phenomenon of eyeshine (see Chap. IX) is even more persuasive and allows safe conclusions about the optical processes occurring in the eye. As we shall see, this also supports the superposition image. In conclusion, anatomical and photomechanical behaviour, as well as the direct observation of the superposition image, leave no doubt as to the eye type of these nocturnal and crepuscular lepidopterans.

3. *Crustaceans*

Despite all efforts, including freezing, I have not succeeded in observing a superposition image in any crustacean. However, this is not surprising, since not only are the crystalline cones not fused with the cornea, but they are also exceedingly soft, in fact almost deliquescent. Many anatomical investigators have already had to contend with these difficulties.

Certainly, some crustaceans have apposition eyes, like that of *Limulus* (see Chap. II), but the eyes of many crustaceans are so similar to those of insects where a superposition image has been demonstrated that it is difficult to conceive of a different mechanism. The similarities include: well-developed crystalline cones, the iris pigment and its characteristic migration, the retina lying at a distance from the dioptrics, the retinal pigment, the tapetum and also the eyeshine which will be treated later.

The crustaceans which I have investigated include: the local river crayfish (*Astacus fluviatilis*), *Palinurus, Palaemon, Peneus membranaceus, Crangon, Scyllarus, Sicyonia sculpta, Pangurus, Nica edulis, Carcinus maenas, Dromia vulgaris, Maja, Pisa, Portunus* and *Galathea*.

Thus representatives of the three decapod families, the brachyurans, macrurans and anomurans, have superposition eyes, whilst *Squilla mantis* (a stomatopod) and, as we saw, *Limulus*, have apposition eyes.

In crustaceans the cornea is much less important as a light-refracting organ than in insects, particularly the beetles. It is thin and the surfaces are only weakly curved. It is clear that the cones play the major role, particularly the strongly refracting portion, which consists of four pieces and has often been likened to a cylindrical lens. In many preparations the refracting surfaces do not appear spherical, there being instead four rounded domes corresponding to each of the four pieces. However, occasionally one also sees fine spherical or hyperbolic end surfaces and since one could hardly suspect such basic design differences in otherwise similar eyes, one must assume that the former appearance is an artefact of swelling.

Although I could not show that this four-piece crystalline cone functions as a lens cylinder in fresh eyes due to their fragility, the results from eyes fixed in alcohol were quite clear. In *Palaemon*, for example, both ends of the lens cylinder are flat (Figs. 51, 52). Nevertheless, even sections of these embedded in damar lacquer project very nice inverted images when their axes are aligned with the microscope. These are similar to those I have described in *Lampyris* after removal of both end surfaces. Further, the rays from one object point on the axis converge into one image point as if the four pieces comprising the cone were a homogeneous mass. The layered construction could also be demonstrated in other ways. Thus if one places the cones in the microrefractometer with a medium refracting either more weakly (water) or more strongly (monobromonaphthalene) than the outer layers of the cone, one sees an exceedingly vivid image of the four-sided prism (this is the shape of these improperly termed "cones"). As expected, in weakly refracting media the prism appears clearly illuminated from the side of the microrefractometer screen with the surface of the other side lying in deep shadow. In strongly refracting media the illumination appears to be reversed. However, if the cones are placed in fluids of intermediate refractive indices, their angular appearance becomes less and less distinct; they appear as round columns illuminated from one side or the other. This can only result from the layering (which functions as a cylinder lens for light perpendicular to the cones' long axis) having a greater optical effect than the external flat surfaces since these are now in contact with a fluid of similar refractive index. Finally, the apparent refractive index of the cone, i.e. that of the fluid in which the cone does not appear to be illuminated from either side, is ca. 1.622. This is impossibly high and the strong refraction can only have arisen through the layered structure characterizing the cone as a lens cylinder.[28]

Using the microrefractometer, a small body can often be seen lying in the interior of the cone. It also has approximately the same shape as the cone and is less refractive than its surroundings. Since not all cones show the phenomenon, I cannot attribute any significance to it and do not exclude the possibility that it is an artefact.

Since lens cylinders also form a major part of the dioptric apparatus in crustaceans, we can probably assume a superposition image even though it has not been seen directly.

It should be emphasized that the formation of a superposition image does not require each dioptric unit to function as an astronomical telescope focused at infinity. It can also be focused at shorter distances. As we saw in the *Limulus* cone, in a lens cylinder of length equal to its focal length, the two principal rays from two points on the object emerge parallel. Accordingly, the cone needs to be only a little bit longer, or to have a convex refractive surface added, for the principal rays to form the inclination to the axis necessary for a superposition image. In this situation the rays from one point on the object could even be slightly divergent. The retinal image will then be less sharp than in the ideal superposition image, but the defect will be largely corrected, at least in the light-adapted eye by the intervention of the iris pigment (see Chap. III). It is possible that at least some of the crustaceans under discussion project such modified superposition images.

The structure of the eye and the migration of the retinal and iris pigment convince me that these crustaceans have superposition eyes, any other possibilities being practically excluded. For example, it might have been considered that such eyes work on a catoptric principle as in *Phronima* (see Chap. VIII). Conceivably the cornea and cone could function in this way, but it is impossible that the transparent parts behind the cone which are fluid (congealing into spherical masses on fixation) and above all of low refractive index, could be capable of guiding light by internal reflection. Indeed the old term "glass body" describes this tissue most aptly in the crustaceans. Thus even though each cone is enclosed in a sheath continuing up to the retina, and although this sheath is not empty behind the cone, the overall effect is of a completely transparent material which, as demanded by the superposition image, lets light through equally in every direction. In *Phronima* the crystalline cones continue up to the retina and stand out so well in the living animal because their refractive index is much higher than that of their surroundings

What about an apposition image, as found in *Limulus*? The large separation between the dioptric apparatus and the retina speaks against it but does not in principle exclude it. But what would be the significance of the extensive, light-induced, migration of the iris pigment when an apposition, or even a catoptric, image were the case? For a catoptric image it would make no sense at all. For the apposition image one could concede a relative dimming of the retinal image with increasing resolution as the pigment moves backwards. The arrangement could hardly be considered functional, since it is clear that in this way many rays emanating from the object point located on the ommatidial axis would also be intercepted. However, on this argument one could accept a form of compromise between superposition and apposition images as in the modification just described.

There is still a final consideration, namely the results of the study with the ophthalmoscope; here the dark-adapted eyes of most crustaceans behave very similarly to those of moths. As described in Chap. IX the eyeshine seen with the ophthalmoscope can only be explained with an eye structure where one image point on the retina is formed from rays which have passed through numerous facets. This is the case only for the superposition image.

Although some decapods have superposition eyes, it cannot be denied that some have eyes showing the anatomical transition to the apposition eye of *Squilla*. These transitional forms and the behaviour of their iris pigment are of special interest.[29]

In the macrurans, the retina is indeed quite remote from the dioptric apparatus, and has a distinct distal boundary, as in the nocturnal insects. Examples include: *Palinurus* (Fig. 50), *Sicyonia* (Figs. 53, 54), *Peneus* (Fig. 47), *Scyllarus*

(Figs. 25, 26) and probably also *Palaemon* (Fig. 52). As we saw, in complete analogy to the moths, pigment migration is also most extensive. In contrast, the retina in many brachyurans consists of long retinulae. The remote retina of the macrurans finds its analogue in proximal, spindle-shaped swellings of the retinulae resting on the basement membrane; e.g. *Portunus* (Figs. 37, 38), *Pisa* (Figs. 48, 49), and *Galathea* (Figs. 45, 46). However, this swelling now only represents a part of the retinal element, correspondingly the pigment migration is diffuse and differs from that found amongst the macrurans.

A further step takes us to other brachyurans where the retinal elements no longer have spindle-shaped swellings but run from the basement membrane to the cones as cylindrical rods, e.g. *Maja* (Figs. 41, 42) and *Carcinus maenas*. Here we have already arrived at eyes similar to those of *Squilla mantis* and the butterflies (Fig. 40) but which still differ from these in exhibiting pigment migration, even though this is no longer very extensive. Indeed, this pigment migration would be incomprehensible if these eyes did not still have the last remnants of a superposition image. First *Squilla*, which no longer belongs to the decapods, has an eye perfectly analogous to that of butterflies, i.e. probably also having a pure apposition image and no longer showing any migration of the iris pigment.

These examples form the basis of my hypothesis of transitional forms between superposition and apposition eyes. This transition seems to be realized within the ranks of the decapods, and the retinulae with their proximal spindle-shaped swellings appear to express the double function of the eye: both as an apposition eye (long retinulae reaching distally), and as a superposition eye (short proximally lying retinulae) depending on the intensity of illumination or the position of the pigment.

The significance of these spindle-shaped swellings will be discussed further in the next chapter where additional support for the view presented here will be given from the behaviour of the retinal pigment.

B. Apposition Images

In *Limulus* the retina is directly attached to the dioptric apparatus (see Chap. II). This is a characteristic feature of apposition eyes. The physiological significance of the dioptric apparatus has been described in detail for *Limulus* and it will suffice here when we find the basic characteristics in other eyes to classify them as apposition eyes, especially when the retina is at the expected level. That the retinulae in most arthropods are longer than in *Limulus* is irrelevant in this respect; the most decisive feature is the direct connection between the retina and the dioptrics.

1. Insects

I cut an eye slice from a fresh bumblebee (*Bombus terrestris*) at such a depth that, when observed in a hanging drop, it appeared completely black apart from regularly arranged small bright spots. These correspond to the tips of the cones and are the apertures through which light can reach the retinulae. With a given position of the microscope mirror only a group of these appear bright. If one moves an opaque object in front of the eye, the bright spots disappear in the sequence expected for an erect retinal image, assuming that each bright spot is detected by one retinal element.

With a point light source the group of bright spots now appears smaller. If the point source is moved the group moves in the sense of an erect image. As in *Limulus* the plane of focus where the spots are smallest and brightest corresponds to the tips of the crystalline cones which, in cross-sections through the eye, are seen to coincide with the narrow apertures in the iris pigment. If one raises the focus each bright point disperses into a blur circle. The axes of these blur circles do not converge appreciably and together they form a large blur circle which also moves in the sense of an erect image when the point source is moved.

The bright spots at the bases of the dioptric apparatuses represent their focal points. Thus there must be a small inverted image at this point, as in *Limulus*. I could not, however, resolve this image because either the apical surface of the cone, or the aperture in the pigment was too small. In contrast the image can be well seen in the fly (*Musca vomitoria*). Note that the image seen by Leeuwenhoek and Gottsche was only the image projected by the corneal facet, since they had brushed out the eye, thus removing all the cones. In my case the cones were carefully preserved in position by freezing.

Returning to the bumblebee, one can observe that a point source is indeed focused at the position described, however, a second point source can be moved almost right up to the first (within 2–3 cm at a distance of 60 cm) without its image appearing next to the first. If the second point source is brought even nearer, the focus from the first point source simply appears to become brighter. In other words, the two poor images had simply blurred into each other.

However, the second point source reveals a phenomenon which is of greater importance. Namely, a second group of bright spots appears next to the first. Thus two object points project two juxtaposed images, each consisting of the brightening of a group of cones, which can be detected by the underlying retinulae. One can bring the point sources very close together before the two groups fuse with each other. These groups of illuminated retinal elements, each corresponding to a different point in the visual field, represent the true retinal image, completely analogous to that in *Limulus*. The function of the dioptric apparatus of each ommatidium is not so much to project an image as to guide as much light as possible to the retinula from points lying on the ommatidial axis and to restrict light coming from other parts of the visual field.

As in *Limulus*, I could also observe the characteristic ray paths in the housefly. An eye slice from a frozen fly also revealed an image at the apical surface of the cone, and when I followed the rays projected from two very closely juxtaposed point sources backwards I saw that the axes of the ray cones did not significantly diverge.

In the fly one can also separate the dioptric function of the cornea and the cone. With two point sources (candle flames) I observed the two groups of illuminated cones and moved the candles together until the two groups started to merge. The cones common to each group now each showed two image points. Their separation was measured as 0.003 – 0.004 mm when the object points were 50 cm from the eye with a mutual separation of 3 cm.[30] I now brushed out the eye and brought the cornea, freed from cone cells, pigment etc. under the same conditions in the hanging drop. Now to my surprise the separation of the two image points was not noticeably changed.* Thus in the fly the position of the focal point is not noticeably influenced by the crystalline cone.

It does not necessarily follow that the cone has no optical function. Presumably its refractive index is higher than that of the haemolymph, and it is attached directly to the back of the cornea. If it had no convergent power, it would thus be expected to shift the image backwards, and this it does not. Nevertheless, the optical function cannot be very significant.[31] This experiment shows that Grenacher's classification of the fly's eye as a pseudocone eye is correct.

A large dragonfly yielded results similar to the fly. The structure of the eyes of these two animals is very similar and one can also expect their dioptric performances to be similar. One again sees the luminous apertures in a good frozen eye slice preparation. One also sees how they disappear and reappear in the sense of an erect image as an object is moved across. A point source illuminates a rather large group of apertures which again moves with the object. A second point source produces a new group, the two groups still do not coincide when the two objects are moved to within a few centimetres of each other (at 60 cm from the eye), and yet, as in the bumblebee, one cannot distinguish the images of the two object points in single ommatidia. Again the pigment aperture is too small. For this reason the experiment of separating the optical function of cone and cornea could not be performed. It is clear that, all else being equal, the groups of ommatidia illuminated by two point sources will be larger the greater the radius of curvature of the eye, and also that they must be further apart from each other. It should be noted that these factors determine the resolution of the eye. In many dragonflies the eye is divided into two regions with different structure; this will be described later in detail (Chap. VII).

Similarly, I was able to demonstrate the characteristics of the apposition image in numerous insects. Amongst the beetles I can cite *Dorcadion aethiops* and one of the click beetles (Elateridae).

In *Limulus* I drew attention to the optical significance of cones lying obliquely to the corneal surface and how this arrangement results in a large visual field despite a flat cornea. I found this characteristic (both the thick corneal cylinders and cones lying obliquely to the corneal surface) again in the hornet (*Vespa crabro*, Fig. 27), which has similarly flat eyes. Since this animal moves around in cracks in dry trees and the like, a strongly curved cornea would be susceptible to damage. As in *Limulus*, this is probably the reason for the development of this apparent defect in the optical design of the eye, which, however, allows for an expanded visual field with a flat cornea.

* The cornea in the fly is described as convex-concave by Ciaccio (1889).

As one sees, all examples of insects with apposition eyes come exclusively from diurnal species.

2. *Crustaceans*

I have already emphasized the great difficulties involved in the investigation of crustacean eyes. As a result, apart from *Limulus*, I can cite only one example where optical investigation of a fresh eye shows the presence of an apposition image, namely *Squilla mantis*.

From their anatomical structure, though, it seems possible that both *Maja* and *Carcinus maenas* also have apposition eyes. In both cases the retinulae are of uniform thickness and extend up to the cone; the interposed pigment would inhibit the formation of a superposition image. Admittedly a weak reaction of the iris pigment can still be seen, which is not easy to understand under the assumption of an apposition image.

In *Squilla mantis* an inverted, albeit unclear image of an external object is produced in each ommatidium. Thus two point sources result in two image points at the base of the cone. The associated rays separate proximally into two sharp dispersion cones with parallel axes as in *Limulus*. I saw this optical effect in both the strongly spherically curved and the almost cylindrically curved parts of the eye.

Microscopical investigation of light- and dark-adapted eyes reveals no obvious pigment migration, further, the retinulae extend up to the crystalline cones, both consistent with an apposition image.

In other respects the eye of *Squilla* is most unusual and will be discussed in detail later (Chap. VIII).

C. *Bi-Functional Eyes*

As already repeatedly elucidated, a number of facts can only be understood on the assumption that some eyes see with a superposition image in the dark and with an apposition image in the light.

Actually the image in *Hydrophilus* or *Lampyris* is already an apposition image in bright sunshine if we assume that the pigment sheath which has migrated from the cone towards the retina is dense enough to restrict the light passing through it to one retinal element. Only, in *Lampyris* and *Hydrophilus* the retinulae first start a considerable distance behind the dioptric apparatus.

Almost all true diurnal species (e.g. flies) have retinulae extending up to the dioptric apparatus. But it has long been known that a number of compound eyes form a compromise between the two types. Thus the retinulae are divided into two parts, a thick, well-developed proximal portion analogous to the retinulae in *Hydrophilus* or *Lampyris*, and a narrow distal part, apparently an anatomical vestige, extending from the spindle-shaped swelling up to the cones. This anatomical vestige is not necessarily non-functional,[32] and indeed, Grenacher has shown that, at least in certain animals,it still contains all the components present in the proximal swelling. However, its form and shape vary greatly. Thus, in *Campharis fusca*, a relative of *Lampyris*, in the best preparations it is a barely visible thread which can hardly still be considered light-sensitive, especially as it is no longer present at all in the firefly. Again in *Palaemon* and relatives it is so degenerated that one can no longer consider it the analogue of the well-formed structures in other crustaceans, but rather as a sheath for the cone. A glance at the illustrations shows the gradation in the degree of development of the distal retinula , which in the following sequence increases from zero to the thickness of the proximal part: Figs. 13, 47, 50, 53, 15, 28, 37, 48, 23, 16, and 58.

It must be added that I know of no eye in which the retinula is divided into two parts and which does not show a photomechanical reaction of the iris pigment. (The converse statement is not true, since there is a photomechanical reaction in some eyes with long, evenly thick retinulae). The significance of pigment migration is perfectly obvious for the superposition eye, where the sensitive layers are remote from the dioptric apparatus, i.e. where the proximal thickenings of the retinulae lie. It is thus extremely likely that these retinular swellings function as the retina as soon as the eye is in the dark-adapted state. The optical processes and apparatus are then exactly the same as in the *Lampyris* eye.

The transition from the thick to the thin portions of the retinula forms a conical surface which is of optical significance. The rays coming from one point on the object should again converge to a point here, themselves forming a cone which should penetrate the retinal element. The conical surface of the retinula is then advantageous since each ray will strike with a smaller angle of incidence than if the retinula were cylindrical. Thus a smaller portion will be reflected than if the conical swelling were absent, which could well be significant for seeing at low light levels. The function of the retinula swelling as a light-receiving retinal element is also well supported by the following. In the transition from the light- to dark-adapted state, not only does the iris pigment migrate as already described, but, as will be described later in more detail, a pigment layer, which in the day just covers the aforementioned conical surface, withdraws leaving it completely free by night. This can be seen very well in *Pisa* (Figs. 48, 49) or in *Portunus* (Figs. 37, 38). It is as if this surface were unsheathed for the reception of light.

For me, the inescapable conclusion is that light which first penetrates the retinula at the swelling will be perceived and that such eyes have a superposition image.

This is the situation in the dark. In the light-adapted state the iris pigment captures most of the rays and the eye receives a retinal image which is certainly very similar to the light-adapted image in *Lampyris*. Apart from the movement of pigment around the swelling, the iris pigment moves proximally so that virtually only axial light can get through. However, along the axis lie, first the thin, and then the thick portion of the retinula. From the foregoing one can hardly doubt that the latter is light-sensitive; but what is the situation with the thin distal part?

Just as the morphological development shows a continuous gradation, the ability of the organ to convert light into nervous excitation most probably varies from zero to the level prevailing in the thick proximal part of the retinula. Presumably the retinula of the apposition eye is not so long without reason, but probably functions over its entire length. Similarly, the mixed retinula may function over its entire length, the thinner part playing a minor role to varying extents in different eyes.

Thus, for example, in a moth's eye in the dark-adapted state (Fig. 28), not only light entering along the axis (be it through the narrow part of the retinula or alongside it) would reach the point on the retinula marked *a*, but also a number of ray bundles which passed through other facets would converge at *a*, enter the retinula and thus lead to light sensation. In the light-adapted state only rays passing through or close by the thin part of the retinula would reach *a*. Those travelling within the thin section of the retinula already produce excitation, whereby it is not excluded, in fact rather probable, that after passing the thinner section they also produce the same effect in the thicker section.

This hypothesis raises a problem, although it is probably only apparent. Namely, if the thin part of the retinula is also light-sensitive then the resolution of the image must suffer in the dark-adapted state since an oblique ray reaching *a* will, in general, have passed through a number of narrow retinula sections and excited them en route. Accordingly, although a sharp physical image will result at *a*, it will be perceived as if it were surrounded by a blur circle. The image will therefore not be physiologically sharp.

I cannot refrain from pointing out the perfect analogy which would then exist with the human eye. In twilight we also have (neglecting the minor difference caused by the pupil diameter) a sharp physical retinal image. However, we sense it as if the image of every point were surrounded by a blur circle. We still recognize black and white, e.g. on paper, quite well, but can no longer read because the physiological blur circles render the shapes of the individual letters unrecognizable.

As already stated, I consider it most likely that the thin portions of the retinulae are also functional in most animals despite the defect in resolution which must necessarily follow. Indeed, later I will discuss how these blur circles could have a functional significance which completely outweighs the disadvantage of poor resolution. However, even if they were non-functional, that would not alter the theory of these eye forms; even then they could still be used in both ways – or better expressed – as in *Lampyris* or *Hydrophilus* there would be a range of intensities to which they could be adapted. Their ability to regulate light flux would only be incomparably greater than that possessed by vertebrates with their pupil.

If we assume the most distal parts of such retinulae to be light-sensitive, then the light-adapted eye receives an apposition image. We usually estimate the size of this retinal image at the level of the most distal part of the retinula. In the dark-adapted state the eye has a superposition image lying at the level of the retinula swellings. Although these two retinal images are, of course, of different size, it is clear that they must be perceived as being the same size since the retinulae which record the images of two object points are the same in both cases.

Animals with these bi-functional eyes include mainly the crepuscular and nocturnal moths and a large proportion of the brachyuran crabs.

Chapter V The Retina, its Pigment and Tapetum

The most important components of the retina and its pigment have already been extensively discussed. I define the retina as the layer of the retinulae as far as the basement membrane, which Leydig calls the "base of the scleral capsule", and through which nerves from the underlying nervous structures, the optic ganglion, project to the retinulae. In certain animals the retina can be very cleanly isolated. For example in the rose chafer, where it shows a precise array of round openings. The space between two openings is ca. half the diameter of one opening.

Berger (1878) defines the retina as all the layers of the retinulae and the optic ganglion. He distinguishes five layers in this retina, which in many animals are in direct contact with the brain, i.e. they are not separated from the brain by an optic nerve.

This terminology has its good reasons, and it may be that we can recognize analogues of certain layers of the vertebrate retina in the structures of the optic ganglion, yet, in the past, the attempt to draw analogies between the compound eye and the vertebrate eye has been the cause of much confusion. We are accustomed to the retina as a nervous layer within the eye, and, according to Leydig, the eye in general is considered as reaching as far as the basement membrane. Thus, despite certain objections, I take the liberty of defining the retina as above, with the nerves from the optic ganglion entering it through the basement membrane.[33]

The nice finding of Grenacher, that the only constant element of vertebrate eyes and compound eyes is the strongly refracting rod (in many arthropods in the form of the rhabdom), raises the question of what is the function of such a rod. Now it is immediately obvious that it must be a device for capturing rays with moderate angles of incidence. Such rays may enter the rod, but can only leave it again at its end; throughout its length escape will be prevented by total internal reflection. If a reflecting body is located at its proximal end the rays will also be channeled back through the rod, or partly through its neighbours in the same fashion. (This can safely be assumed from the investigations of Brücke in vertebrate eyes with tapeta). Many arthropods also have a tapetum, and in these eyes rays are also incident on the retina approximately parallel to the axes of the retinulae or rhabdoms.

This arrangement suggests that the most fundamental visual process – the transduction of light energy into nervous excitation, be it by photochemical or other means – takes place, or is at least initiated, during the passage of light along the strongly refracting rods. The connection between the rods and the nerve fibres is subserved by the retinula cells (Grenacher), which together with the rods form the retinulae. The often considerable length of the rods suggests that it is not irrelevant for the visual process whether a captured ray takes a short or a long path through the rod. Thus rays passing the rod obliquely and leaving it again may elicit a much weaker excitation than those running the whole length (cf. what is said on p. 63 about blur circles).

In the course of my own rather extensive investigations of arthropod eyes, I have made no observation which did not fully agree with Grenacher's interpretation of the retinulae. Namely, that the retinula is composed of (usually) seven elongated cells which secrete the rods as a sort of cuticular formation along their common contacting surfaces. When these are fused they form the rhabdom, which, in cross-section, thus appears as a seven-armed star surrounded by the unmodified protoplasm of the retinula cells (Figs. 37a, b; 48d; 50d). The connection of these retinula cells, of course usually recognized together as the retinula, with the nerves entering through the basement membrane can be very clearly seen in many animals, e.g. *Maja*. In this crab (fixed in Müller's solution and exposed to the sun before its death) the pigmented nerves merge directly into similarly pigmented retinulae (Figs. 41, 42; cf. also Figs. 37, 38 and Figs. 45, 46).

Recent investigations by Watase (1890) have shown that the rhabdom is laid down by the retinula cells in the same way as the cornea and cone are formed as cuticular deposits by ectodermal cells. Since the retinula cells are considered to be ectodermal sense cells, everything that functions dioptrically or catoptrically in the ommatidium, even at the depth of the retinulae, can be regarded as cuticular formations.[34]

A. The Tapetum

As in the vertebrate eye, tapeta are also found in compound eyes, and as in vertebrate eyes the optical function, namely reflection, can be achieved in two ways. Certain vertebrates (herbivores) have a tapetum made out of fibres whilst others (carnivores, fishes) have one composed of cells. Arthropods fall into those with a tapetum made out of numerous tracheae (insects) and those with a tapetum made out of a strongly reflecting granular substance which is probably stored in cells. The latter is found in crustaceans and to my knowledge has not previously been recognized as a tapetum. Finally, as in vertebrates, many eyes have no tapetum.

By tapetum is understood a layer which reflects light that has already passed through the sensitive layers of the retina so that it can be used twice, thus increasing retinal excitation. Accordingly, tapeta are usually found in nocturnal animals.

In all known cases the same optical principle is in operation, namely that according to which glass powder, soap suds etc. appear white, although their constituents are transparent. This is because light is refracted at every boundary between media of different refractive indices, and the greater the difference the more light will be reflected. For example, if a light ray strikes a glass splinter, part of it will be reflected at the interface between air and glass, another part enters the glass and undergoes another reflection at the surface where it should exit. This can even be total with an appropriate angle of incidence, i.e., no light will emerge from the other side. If some does emerge, it will suffer the same fate with the next glass splinter, so that in the end all the light is reflected and none succeeds in passing through the glass powder. An observer on the illuminated side receives the reflected light and thus sees the powder as white; on the other side he would receive no light and the powder would appear black. This is the reason why white hairs, tracheae etc. appear white in incident light and black in transmitted light under the microscope. It also follows that light reflected in this way travels in many different directions. The powder appears uniformly white even when one moves one's eyes to and fro; if the light was reflected in a particular direction this would not be so but the powder would appear reflective. Reflected light from a tapetum is always diffuse.[35]

With this principle of reflection the shape of the finely distributed particles is only of secondary importance, whether they be round as in crustaceans, hexagonal as in certain fishes, or whether they are long threads as in the fibrous tapeta of herbivorous mammals or the tracheae of insects.

As mentioned, Leydig (1864a) was the first to describe the tapeta in insects. He describes perfectly correctly how, on dissection of, e.g., a moth, one is surprised by the beautiful shining membrane lying in the depth of the eye which is instantly recognizable as a tapetum. I verified his assertion that this derives from small tracheal bundles which penetrate the basement membrane and break up into bundles of very fine tracheae. Each bundle ensheaths a retinula, basket fashion, starting from the proximal end. As Schultze describes, the tracheae end at the level of the transition from the thick to the thin portion of the retinula. Leydig (1864b) illustrates them as ending more proximally. Remarkably, the thought that this could be a tapetum does not seem to have occurred to Schultze, he supposes rather that the tracheae could be a substitute for pigment. Certainly the tracheae will also prevent the spread of light, but in such a way that it will again benefit the retinula through which it passed.

Figures 10 and 11 show the tapetum in the retina of a moth drawn in both transmitted and incident light, and Fig. 12 two tracheal bundles, each belonging to one retinula.

When one considers that the retinula lying within such a bundle is in the incident light position with respect to rays that have passed through it, but in the transmitted light position for rays which originally passed through other retinulae, then the value of this type of tapetum is fully clear, especially when one considers that the rhabdom is also a device for capturing the reflected light. There can be little doubt that this type of tapetum is probably the most perfect found in the entire animal kingdom, not only because of its relation to the retinulae but also because of the use of air as a medium of very low refractive index.[36]

I only know of this sort of tapetum, where just the proximal swollen end of the retinula is ensheathed by the tracheal mass (i.e., where it must be most important for reflection), in nocturnal insects. However, in discussing the tapetum, Leydig (1864a) explicitly emphasizes that he has seen a white glimmer in diurnal animals as well (*Argynnis, Volucella*). Possibly this originates from another form of tracheal ending which tends to occur in diurnal species.

Leydig was also the first to see the thick, sausage-shaped tracheae in butterflies which run between the retinulae, ending distally, like fingers, at the level of the crystalline cones. Later, Thompson Lowne (1884) also described similar tracheae in syrphids encircling each retinula. My drawings from a butterfly (Fig. 35) and a beetle (Fig. 24) show these tracheae as round or polygonal fields between the retinulae. Schultze also confirms this type of trachea, stating, however, that only four run along each retinula in butterflies. He also saw such blind-ending tracheae in flies and dragonflies.

I have no doubt that these relatively thick, distal tracheae are morphologically closely related to the tracheal tapetum previously described, but I believe their functional significance to be quite different. What is the point of a tapetum in the distal part of the eye? It is obvious that any ray travelling within the retinula must be reflected back by these surrounding tubes of air, and thereby after one or more reflections it will end up in the rhabdom where it will serve its visual function. But since the rays undergo these reflections on their way down the retinula, the function cannot be considered analogous to that of a tapetum.

Therefore, such tracheae belong functionally to the actual retina and are discussed here only on account of their morphological relationship to the tapetum.

Unfortunately, one can no longer see the finer tracheae in preparations embedded in the highly refractive resin, which explains why my illustrations do not show the insect tapetum in more detail.

In general, the granular substance which makes up the tapetum in crustaceans occupies the same position as the tracheal bundles in insects and ensheathes the proximal ends of the retinulae. Presumably it lies in cells, though I could not usually convince myself of this (cf. Fig. 48d). In fresh preparations it appears to be composed of round granules with diameter, e.g. in *Pisa*, of ca. 0.001 mm. In transmitted light it forms a black mass which can only be distinguished with practice from the black pigment by the nature of the granules. However, when viewed in incident light it appears bright and shiny yellow. Examples are shown in Plate V. The granules of this tapetum must have an exceptionally high refractive index since the strong reflectivity is not noticeably reduced even in fixed eyes embedded in damar. Indeed, in *Palaemon*, the tapetum still reflected strongly after I had impregnated it with a mixture of aniline and monobromonaphthalene which had a refractive index significantly greater than 1.6. This leads one to suppose that quite exceptional optical behaviour prevails here.[37] In crustaceans one has the advantage that, even in properly embedded sections, one can always demonstrate the tapetum, which is not the case in insects, where the tracheae lose their air and become filled with damar. The chemical nature of the tapetum remains to be established. At the moment I can only state that it is insoluble in all the fixatives and preservatives I used (i.e. picric acid, chromic acid, mercuric chloride, Müller's solution, alcohol, ether, xylene, water, weak formic acid and strongly alkaline soap solution).

One of the most beautiful and astounding images one can see under the microscope is when one views a radial section of a crustacean eye. Under normal conditions one sees what appear to be black layers of retinal pigment. On blocking out the transmitted light, in place of the presumed pigment one sees a glistening gold layer (Fig. 39). For the best effect one should use low magnification or a simple microscope, as otherwise the objective blocks out too much of the incident light.

At higher magnifications, and on thin sections, one can see that the tapetum forms a layer adjoining the basement membrane, completely covering the proximal ends of the retinulae. Distally it terminates in sharp or club-shaped processes projecting between the retinulae.

In cross-sections of the retina, most clearly in the dark-adapted eye of *Palaemon*, one sees the round rhabdom in the centre of the, also rather circular retinula, and the tapetum directly attached to this in a continuous mass. This is, of course, not the case in the distal part of the retinula, where the tapetum no longer forms a continuous mass (cf. Figs. 48c, 49f, 50d).

The physiological significance of this is quite obvious. I will return to these retinula endings again when discussing the retinal pigment.

Apart from this tapetal layer at the proximal ends of the retinulae, and, of course, apart from the layer previously designated as the iris tapetum, in many crustaceans one also finds a second tapetal layer which usually lies outside the eye proper in the optic ganglion (*t* in Figs. 51–54, see also Figs. 47, 49). It consists of numerous matted strands which may well be cell processes. Their arrangement is reminiscent of the pigment distribution in the stroma of the human choroid.

These two layers tend not to be rigorously divided but are frequently linked together by irregular fibrous and tuberous connections. These pass through the basement membrane, thus giving the impression that the layer in the ganglion might be a reserve layer from which tapetal substance could migrate into the eye (cf. Figs. 48, 49).

I found a tapetum in almost every macruran investigated; in the crayfish (*Astacus fluviatilis*), where it surrounds the retinulae for about half their length, in *Crangon*, in *Palaemon* which has two layers (Figs. 51, 52), in *Palinurus* with one layer (Fig. 50), in *Nica edulis*, in *Sicyonia sculpta* (Figs. 53, 54, 39) and *Peneus membranaceus* (Fig. 47). The last three all show a double tapetal layer. Only in *Scyllarus* could I find no tapetum. It should be recalled that the macrurans have the most unquestionable superposition eyes, and also the most pronounced pigment migration. These thus appear to be the nocturnal species amongst the crustaceans, and this is correlated with the presence of a tapetum.

Amongst the brachyurans I found a tapetum in *Carcinus maenas, Dromia vulgaris* (Figs. 55, 56) and *Galathea* (Figs. 45, 46) whilst in *Maja* (Figs. 41, 42), *Portunus* (Fig. 37) and *Pagurus* none could be recognized. Neither did I find one in *Squilla*. It must be emphasized that the absence of a tapetum in a preserved eye (and most animals were preserved) does not prove its absence in vivo. In certain animals, the tapetum might be of such a nature that it is destroyed or rendered invisible by the preparation. For example, in my preparations of *Pisa* there is a weak but noticeable reflection so that, for a while, I was in doubt as to whether there was a tapetum or not.

It is remarkable that, at least as far as I know, no one has previously recognized the retinal tapetum in crustaceans. I know of only one statement in the literature indicating that it has been seen, though not recognized as such. Thus Grenacher (1879, p. 119) writes, in his description of the rhabdom in *Mysis*:

"The clearly demarcated and moderately sharp, proximal end is barely discernible because it is surrounded by processes of a characteristic earthy pigment which ascends with the nerve fibres of the optic ganlion behind the basement membrane and which unfortunately cannot be removed by aqua regia".

An accompanying illustration left me in almost no doubt that, in this, Grenacher had seen the tapetum, although I cannot say this with certainty as I have not investigated *Mysis*. It is drawn in a yellow tone, i.e. more or less as one would see it in incident light. Furthermore, Schultze illustrates the proximal ends of the crayfish retinulae surrounded by a yellow granular pigment, but without further comment.

B. The Retinal Pigment[38]

The retinal pigment has long been known as the proximal mass of pigment and it has often been stressed that it more or less surrounds the retinulae. Leydig (1864a) also recognized that pigment frequently lies behind the basement membrane. He writes (p. 29) : "The black pigment does not always first appear within the eye capsule (*sclerotica*), but the bundles of nerve fibres coming from the optic lobes are also very often more or less strongly pigmented. The pigment accumulates at the point where the nerves enter the eye capsule, whereby a thick dark zone can develop behind the basement membrane, as for example in *Acherontia atropos*. The pigment then lies within the neural matrix (Neurilem)."

I have nothing fundamental to add to the available descriptions of this retinal pigment as far as insects are concerned. In all those eyes which I have designated as typical diurnal eyes there is no sharp boundary between the retinal and the iris pigment; the pigment is particularly dense at the proximal and distal ends of the uniformly thick retinulae. Distally, it ensheaths the connection between retinula and crystalline cone and proximally the connection with the nerves. It tends to be most sparse in the middle of the retinula. (cf. Figs. 32, 40).

As Leydig correctly describes, the amount of pigment found in the optic ganglion differs widely between species. This is best seen in the illustrations (see Figs. 13–17).

It must be emphasized that the pigment in nocturnal eyes differs significantly from that in diurnal eyes – at least among the lepidopterans. In moths, most of the pigment is in the form of short pyramids resting on the basement membrane, with their ends projecting between the retinulae (Figs. 30, 31). I saw this in all the true nocturnal moths I investigated. The humming-bird hawkmoth (*Makroglossa* Fig. 17), which flies in bright sunshine, was the only exception. In this species long strands of pigment stretch distally from the basement membrane. They lie between the thick portions of the retinulae and end at the level of the retinula cell nuclei.

More retinal pigment is found in the eyes of nocturnal beetles, e.g. *Hydrophilus, Lampyris* and *Colymbetes*, where the proximal ends of the retinulae are completely ensheathed (Figs. 1, 13, 16).

Here the rose chafer (*Cetonia*) and relatives occupy a position similar to that of the humming-bird hawkmoth amongst the moths. The retinal pigment forms long strands reaching as far as the thick portions of the retinulae (Fig. 23). In beetles, as in moths, the retinal pigment is sharply separated from the iris pigment. As shown by observations on *Hydrophilus, Lampyris* etc., this separation is maintained even when the iris pigment is in the extreme light-adapted position.

I am unable to say if the black pigment is contained within cells. It seems not to be; however, my experience with vertebrates shows how easily one can be deceived in this. Further, it is almost inconceivable that there are no cells involved in pigment migration – what else could provide the motile force?

C. The Photomechanical Response of the Retinal Pigment

I have not succeeded in demonstrating a mechanical action of light on the retinal pigment in insects.[39] In crustaceans the action is quite obvious and is best discussed along with a description of the retinal pigment itself.

I start with *Palaemon* (Figs. 51, 52), a widely distributed macruran with eyes that can be easily sectioned and are therefore to be recommended for future investigations.

In the dark-adapted eye one can see the two tapetal layers already described (Fig. 51), and in between them, i.e. directly proximal to the basement membrane, the black pigment layer. Cross-sections through the proximal ends of the retinulae reveal elegant transparent stars, the retinulae, surrounded by tapetum. In the twilight-adapted eye* part of the pigment has disappeared from its original position, i.e. the pigment layer behind the basement membrane is no longer so dense, and now there are considerable quantities of pigment between the retinulae, particularly their distal ends. However, the tapetum surrounding the proximal ends of the retinulae is still clearly visible. Single tapetal strands project distally, passing through both the retinal and the iris pigment to make contact with the iris tapetum. In the light-adapted eye (Fig. 52) the retinulae are richly coated in pigment over their entire length. The zone proximal to the basement membrane, which originally contained the pigment, now contains only sparse remnants; instead, abundant tapetal material fuses with the more proximal tapetal layer so that the latter appears to fade out distally.

It is very difficult to say whether these different appearances arise solely from the movement of pigment or whether the tapetum is also mobile. Initially I believed the latter, but more careful observations led me to the conclusion that only the pigment moves and the different appearances of the tapetal material merely express the degree of concealment by the pigment. For example, the apparent absence of tapetum in the most distal layers of the optic ganglion in the dark-adapted eye is probably due to the pigment absorbing the light before it

* This is the eye of an animal killed during the day in the back of a room; a light-adapted eye is the eye of an animal killed in direct sunlight.

reaches the granules of the tapetum. Similarly, the tapetal layer at the proximal ends of the retinulae seems to disappear completely in some light-adapted eyes. This is probably because the pigment has not migrated as far distally as in the eye shown in Fig. 52.

For an understanding of the optical function it is of particular importance to identify the space occupied by the pigment during its distal migration. I mentioned that in proximal cross-sections of the dark-adapted eye the star-shaped retinula is directly surrounded by tapetum. In the twilight-adapted eye one now sees pigment within the star, filling the arms so that from the whole star only the highly refractive centre, the rhabdom, remains transparent (cf. an analogous picture from *Pisa*, Fig. 49f). For this one needs to look at rather thin cross-sections, alternately in incident and transmitted light; in the latter case one sees no trace of the star and only the bright rhabdom; in the former case one sees a dark star surrounded by a bright tapetum, similar to Fig. 50d except that one has to imagine a bright rhabdom in the centre. With careful control of the illumination one can convince oneself without any doubt of the presence of the pigment granules within the arms of the star. Indeed, in *Carcinus* I found certain places where the pigment had not yet permeated the entire arm, unpigmented zones being left free in club-like swellings of the arms. (One must imagine, in Fig. 50d, that both horizontal arms are more strongly swollen and transparent in the middle). I saw two to four such incompletely pigmented arms in one retinula.

It has generally been considered inconceivable that pigment enters the retinula itself, and very improbable that the retinula has a star-shaped cross-section at its proximal end. The above interpretation may have to undergo revision when the crustacean eye is investigated in more detail, but for the moment it seems to be the most reasonable interpretation. Thus the strongly refracting glassy centre of the star must be the rhabdom, and the surrounding tissue has a star-shaped border. Now it may be that this is only part of what is commonly called the retinula, but then the retinula would have to contain not only pigment in its central part, but also tapetal material in its peripheral part.[40]

Whatever the interpretation in morphological terms, it is clear that in the dark-adapted eye light, after passing through the retinula and rhabdom, will hit the tapetum and thus be reflected, whilst in the twilight eye, pigment has already moved between the incident light and the tapetum, and reflection will thus be hindered.

Apparently this is not the only route of pigment migration. Rather, the majority appears to migrate between the retinulae and associated tapetum so as to ensheath the distal halves of the retinulae, which are free of tapetum. In the light-adapted eye the pigment is especially dense right at the level of the distal ends of the retinulae. However, there always remains a free zone between this and the iris pigment which has migrated proximally.

If one looks at a dark- and a light-adapted eye (Figs. 51, 52) with a simple microscope one gains the impression that in the dark-adapted eye there are two tapetal layers separated by one pigment layer, whilst in the light-adapted eye there are two proximal tapetal layers and one distal pigment layer. In the latter case the two tapetal layers often appear to be fused into one, and the pigment now occupies a zone which before contained neither tapetum nor pigment.

I observed the described behaviour of the pigment in dark-, twilight- and light-adapted eyes in a series of specimens from two species of *Palaemon*. Although I am convinced that the description is correct, in one dark-adapted eye some pigment lay distal to the tapetum over two thirds of the ommatidia. I have also noticed such irregularities in other crustaceans. Since the photomechanical action of the retinal pigment is easily and confidently demonstrable, I explain the cases where the pigment did not assume the expected extreme position in that it was almost always necessary to let a little light into the darkened cage or room in order to catch the animals before killing them. In other cases the animals were in a small container from which I could pour out the water in the dark and replace it with alcohol. It seems to me probable that this, albeit short and very moderate, illumination was sufficient to influence the results in those few cases. Naturally, I always saw migration in the sense described, only with respect to the extreme dark- and light-adapted positions can one talk of irregularities. The state of health of the animals can also affect the results. Thus, as I have already mentioned, the photomechanical action of the iris pigment cannot be demonstrated in weak or dying animals.

I also investigated the dark-, twilight- and light-adapted eyes of the spiny lobster, *Palinurus*. In the dark, the retinal pigment lay proximal to the retinal tapetum (right half of Fig. 50). However, over large regions of this eye some black pigment also lay distally from the tapetum (left half of Fig. 50), probably because it required some time in the light to catch and decapitate the animal, which was large and defended itself vigorously. In the twilight-adapted eye, this was already the case all over the eye and the distal ends of the retinulae were more or less invisible as they were ensheathed by pigment. Proximal to this pigment came the tapetal layer, which was also invaded by numerous strands of pigment. It is noticeable that, both in the dark- and the twilight-adapted eye, all the pigment is found in the layers of the retinulae ; there is no trace in the optic ganglion. As in *Palaemon* the cross-sections of the retinulae are seven-armed stars, and in the twilight-adapted eye the arms of the stars contain pigment, with the tapetal material outside (Fig. 50d). In the dark-adapted eye some pigment could also be seen in the most proximal part of the retina, but this is probably because it was not completely dark-adapted.

I saw very beautiful and extensive pigment migration in *Sicyonia sculpta*. The retinulae are clearly striated and a rather thick basement membrane separates them from the fenestrated layer of the optic ganglion containing the nerve branches (Figs. 53, 54, 39). The more central layers of the optic ganglia are nicely compartmentalized. In the dark-adapted eye all the retinal pigment (*r.p.*) lies in the fenestrated layer. The most distal layer of the tapetum (*r.t.*) ensheaths the proximal ends of the retinulae, and the proximal layer (*t*) lies within the vicinity of the plexiform layer. In both the twilight- and the light-adapted eyes the fenestrated layer is completely free of pigment, and also contains no tapetum. The pigment has migrated through the distal tapetal layer and ensheathes the distal ends of the retinulae. In migrating distally the retinal pigment moves right up to the iris pigment (*i.p.*) which has moved extensively in the opposite direction. In this animal there appears to be no pigment at all left in the distal tapetal layer so that, in contrast to *Palaemon*, light is then not absorbed in the proximal end of the retinula. However, the greatest part will already have been absorbed distally before it enters the retinula. Probably the only ray bundle to reach the retinula intact is that from the associated crystalline cone. Of course, the same can also be assumed for *Palaemon*, where the pigment ensheaths the retinulae both distally and proximally when light-adapted.

It should not be forgotten that in these crustaceans there is a pigment-free zone stretching from the crystalline cone to the retinula. Schultze, who considers this zone part of the crystalline cone, described how it envelops the distal end of the retinula. Thus neither the iris nor the retinal pigment, even when the latter is densely packed at the distal end of the retinula, can completely block out the light bundle from the overlying crystalline cone. Both types of pigment appear to complement each other in converting the superposition image into a much less intense apposition image similar to that seen in diurnal species.

Nica edulis behaves very similarly to *Sicyonia* and gave an equally clear result. In the dark the retinal pigment again lies behind the basement membrane amongst tapetal processes. However, most of the tapetum is again found in two layers, one between the proximal ends of the retinulae and the other proximal to the pigment layer in the optic ganglion. In the light-adapted eye the pigment has disappeared from the ganglion and is found characteristically mingled with the tapetum, mainly between the distal portions of the retinulae.

I could not demonstrate a tapetum in *Scyllarus*. The retinal pigment behaves similarly to that in brachyurans. In the dark it is accumulated around the proximal ends of the retinulae and moves to their distal ends in the light.

I only succeeded in demonstrating migration of the retinal pigment in the crayfish after many fruitless attempts, showing once again that the animal must be in good condition for the experiment to succeed. The first success came after I had kept the animals well-fed in a tank with flowing water for several days after their purchase. A number of these were then killed in alcohol in the dark of the night, and a number after they had been exposed to bright sunlight.

The crayfish has only one clear tapetal layer, which lies between the proximal ends of the retinulae. In the dark the pigment lies in the fenestrated layer. In the light the majority has left this layer and now densely sheaths the distal parts of the retinulae.

As mentioned above, the specimens of *Peneus membranaceus* which I investigated had more or less no pigment in the retina, but the usual two tapetal layers were present. The problem of tapetal migration can thus be tackled without the interference of pigment migration. Accordingly, I sectioned several eyes of this crustacean and found the tapeta to be identical in the dark- and twilight-adapted eyes. In one light-adapted eye the most proximal tapetal layer had vanished (*t* in Fig. 47). I will not venture to conclude from this finding that tapetal masses can also migrate distally, but perhaps it will motivate some researcher who has living specimens of these beautiful crustaceans at his disposal to answer the question of tapetal migration. It is also possible that I was dealing with an individual peculiarity.[41]

The situation in the brachyurans is somewhat different to that in the macrurans. The pigment tends not to appear in such sharply defined layers, and the movements have rather the character of a shift of the main body in one direction or the other. Where the retinulae adjoin the crystalline cones, the proximally migrating iris pigment mingles with the distally migrating retinal pigment, obscuring the boundary between them. The iris pigment then often lies around the distal ends of the retinulae (Figs. 49, 56) so that it could be mistaken for retinal pigment. However, since the pigments are very clearly separated in the dark I believe one must stick to the terminology and the distinction of the two pigment types.

Radial sections through dark-, twilight- and light-adapted eyes of *Dromia vulgaris* (Figs. 55, 56) showed the following: in the dark the proximal ends of the retinulae are surrounded by a tapetal layer, distally they are free of both tapetum and pigment. In one eye there was some pigment amongst the tapetum, in another no pigment was present at all in this layer. The pigment is rather richly accumulated just beneath the basement membrane in the optic ganglion. In both the twilight- and the light-adapted eye the pigment has disappeared and is found in a dense layer surrounding the swollen portions of the retinulae.

Pagurus has no demonstrable tapetum, but even in the dark-adapted eye there is pigment lying between the proximal ends of the retinulae. Pigment is also found in abundance in the adjoining parts of the optic ganglion. In the light the pigment has again almost completely disappeared from the latter position. Instead, the swellings of the retinulae are richly covered in pigment both proximally and distally. The pigment did not move so far distally as to fuse with the iris pigment.

In *Maja* (Figs. 41, 42) I could not confidently demonstrate any migration of the retinal pigment.

Pisa (Figs. 48, 49) has a layer of weakly reflecting tapetal material at the proximal ends of the retinulae. It is disposed exactly as the distal tapetum in other crustaceans. In the dark the pigment again lies in the fenestrated layer and only projects sparsely through the basement membrane into the retina. In the light it has migrated to ensheath the swellings of the retinulae. In doing so it develops into slender strands, seven to each retinula. Since each strand is discontinuous it forms a row of dark points. A cross-section of the retinulae at the level of the transition from the thick to the thin portion shows a seven-armed star (*e*).*

More proximally (*f*) the pigment is substantially reduced and surrounded by tapetum. Perhaps due to the weakly reflecting properties of the tapetum, in this species one can actually recognize the tapetum as being contained within cells with nuclei (Fig. 48d).

Portunus behaves very similarly (Figs. 37, 38), except that I could not recognize any tapetum; instead, even in the dark, the proximal half of the spindle-shaped swellings of the retinula is covered with pigment (Fig. 37a, b). Again the retinula reveals a seven-armed star in cross-section with the translucent rhabdom in the centre. Pigment is again found abundantly in the fenestrated layer. In the light most of this migrates into the retina, where it is now densely packed around the distal halves of the spindle-shaped swellings. Here it is more densely packed than it was around the proximal halves in the dark.

* The figure is imperfectly reproduced in the plate. the arms being formed of pigment, with a transparent rhabdom in the centre.

Galathea (Figs. 45, 46) has a tapetum between the proximal ends of the spindle-shaped swellings. Even in the dark I also found quite a lot of pigment between the retinulae and the tapetum. Possibly this again is a result of a brief illumination before death. There is also copious pigment behind the basement membrane in the dark. In the light this is almost all gone, whilst the distal halves of the swellings are richly invested in pigment. The pigment from the proximal ends of the retinulae also migrates distally so that this layer is now dominated by the tapetum alone. The iris and retinal pigment again intermingle in this eye on illumination and together ensheath the entire extent of the retinula with the exception of the most proximal ends.

Finally, in *Squilla mantis* I found no tapetum and also no migration of the retinal pigment.

As most beautifully seen in the macrurans, the majority of the decapods possess a granular tapetum which encloses the proximal ends of the retinulae and functions as a device for reflection. In many species this reflection is reduced upon illumination by black pigment which migrates distally between the retinula and the tapetal material. Even when this is not the case most of the pigment comes to ensheath the distal part of the retinulae thus shielding them from light. In general this pigment comes from the most distal layers of the optic ganglion.

Chapter VI Eyes with Non-Uniform Construction

We have already seen that different eye regions can be differently constructed. In particular the ommatidia at the edge of the eye may have a different inclination to the corneal surface than those in the centre of the eye. This difference, though, is minor and we have already recognized its functional significance. Certain animals, show much more striking regional differences in eye construction and these will now be considered with respect to their optical function.[42]

A frequent occurrence, particularly amongst the crustaceans (amongst the insects notably in the dipterans), is a very marked variation in the curvature of the cornea. Thus the cornea in the middle of the eye may have a far greater radius of curvature than at the edge; the eye has strongly curved edges.

We have already seen that the size of the retinal image (both apposition and superposition) depends upon the radius of curvature of the eye. Thus in such eyes a given object will project a larger image in the centre of the eye and the retinal image of the whole eye will be distorted, peripheral parts of the image, in general, being contracted in directions towards the centre. Eyes with different curvatures in different planes, which are common in both insects and crustaceans, must also show similar distortions in the retinal image. For example *Squilla*, apart from a ring-shaped indentation, has practically cylindrical eyes with rounded ends (see Chap. VIII). Similar, though less pronounced, differences are of common occurrence. Whilst lepidopterans, particularly moths, and some macrurans (e.g. *Palinurus, Homarus* and *Palaemon*), have approximately spherical eyes, the eyes of most beetles, dipterans, brachyurans and many others are markedly non-spherical. They thus all have retinal images which are geometrically dissimilar to the projection of the visual field. The physiological significance of this is discussed later (Chap. X).

In certain dragonfly species (e.g. *Cordulegaster*) the dorsal part of the eye differs from the ventral and lateral parts. As can be instantly recognized from the size of the reflections, the curvatures are different. The colour and, above all, the texture differ substantially. Whilst, in the ventral and lateral parts, one sees the optical phenomena of the pseudopupil (see Chap. IX), there is no sign of this in the dorsal part (Figs. 66, 67).

In sections through the eye, one sees (in large animals even with the naked eye) that the layers of the eye (i.e. from cornea to basement membrane) are thicker in the dorsal eye than elsewhere, that the transition is rather abrupt and that the colouring is different. Under the microscope (Figs. 58, 59) each ommatidium in the ventral eye is seen to be smaller in all its components than in the dorsal eye, though it is relatively longer. Further, in the ventral eye the proximal part of the cone and the entire retinula are blackly pigmented with only the distal part of the cone being surrounded by coloured pigment, whilst in the dorsal eye there is only coloured pigment. The transition is so sudden that next to a typically black pigmented ommatidium there is perhaps one with less black pigment but then the next one already has none at all. With respect to the dimensions the transition is a little less abrupt.

Libellula depressa also shows these regional differences. The size of the corneal facets measured on a brushed out eye was 0.059 mm dorsally and 0.033 mm ventrally (between opposite corners of the hexagons). The distance between the corneal surface and the tip of the cone in *Libellula vulgata* was 0.216 mm dorsally and 0.133 mm ventrally. The maximum width of the cone was 0.058 mm dorsally and 0.023 mm ventrally.

Notthaft (1880) also emphasized the difference in facet size and coloration, apparently without having made microscope sections. He also pointed out the abrupt transition between the two eye regions and gave measurements of the facet size. For *Libellula* sp. he quotes 0.054 and 0.028 mm, and in *Aeschna* 0.06 and 0.035 mm. He also found analogous differences in the dipteran *Stratiomys* and remarked:

"As far as I am aware, facets of two sizes lying next to each other in the same eye only occur in predatory insects; furthermore the more finely facetted pattern, enabling a sharper image, is almost always restricted to the ventral and frontal parts of the eye. This leads to the supposition that this organization is closely related to prey chasing and capture..." This remark warrants further discussion below.[43]

In some *Libellulidae* the difference in construction of the two eye regions is already revealed by the coloration, e.g. dorsally dark blue and ventrally light green. In grasshoppers as well, e.g. *Psophus stridulus*, the eye may be divided into two differently coloured halves separated by a sharp line. In dipterans the situation is similar to that in dragonflies. The corneal facets in many species are larger dorsally and often the two regions are also differently coloured, although the separation is not as sharp as in dragonflies. Girschner (1887) has studied dipteran eye pigmentation in more detail.

The curious eyes of *Phronima* show exceptional regional differences and are treated extensively in Chapter VIII. The two parts of the eye are so different that one can justly talk of two eyes on each side, especially as the retina is correspondingly sub-divided. The cones of the dorsal eye are many times longer than in the lateral eye. Here again the dorsal part of the eye shows the greatest deviation from the normal. Thus both *Phronima* and dragonflies have an organ specialized for observing what goes on above them.

I know virtually nothing about the life-style of the *Phronimidae*, but for the dragonfly one can readily attempt to interpret the significance of the two eye regions. The larger dragonflies have a preference for settling horizontally, at least whilst they are active during the day. They also settle preferentially low down on the ground or on low plants. Only during courtship have I seen them flying about between trees high in the air. Thus in the usual situation when a dragonfly is resting on a path and one tries to creep up on it and catch it, then it sees the enemy or the net with the dorsal part of the eye. Also if it sees another dragonfly or insect which it then chases out of love or hunger, then the target is again recognized with the dorsal eye. Once it has caught the insect and is ready to eat it, it sees its prey with the ventral eye, whilst the dorsal eye monitors the proximity of danger or new prey. In a previous publication (Exner 1875), I pointed out that, in general, insects notice or recognize their enemies and probably also friends and prey, from their movement. The resting dragonfly must always notice moving objects with the dorsal eye region, whereas the ventral region is concerned with still objects.

Thus I consider that the reason for these regional differences is that one design is more suitable for the detection of motion and the other for the perception of shapes of still objects. How that is possible will be discussed later (Chap. X).[44]

Chapter VII Short Descriptions of Selected Eyes of Insects and Crustaceans

A. Insects

1. Butterflies

Pieris rapae (cabbage white Figs. 34, 35, 61). A typical diurnal, apposition eye. The bases of the cones are significantly smaller than the corneal facets. Distally, the very short cone is embedded in iris tapetum and proximally in iris pigment which extends as far as the retinal pigment in strands surrounding the tracheae. A cross-section through the retinula (Fig. 35a) shows four central rods,[45] which are not fused into a rhabdom, surrounded by the thick tracheae described by Leydig.

The eye of *Melanargia* (marbled white) is very similar to that of the cabbage white, with iris tapetum, four strongly refracting rods in the retinula neatly arranged in a square in cross-section etc.

The tortoiseshell (*Vanessa*, Fig. 36) has a typical diurnal eye similar to that of the cabbage white and other butterflies. As a peculiarity, even with the naked eye, the cornea is seen to be richly covered in long hairs (*a*, Fig. 36). Thanks to their radial disposition they should hardly disturb the retinal image and also allow the pseudopupils to be well seen.

Epinephele (Figs. 32, 64) is also very similar to the cabbage white. The rhabdom can be seen throughout the length of the retinula. Isolated nuclei are found at the proximal ends of the retinulae. In some sections the basement membrane is isolated and apparently packed with pigment granules. Proximal to this lie characteristic club-like structures.[46] They merge distally into the retinulae, and proximally they appear to make connections with the pigmented nerves. They correspond to the retina of Thompson Lowne.

Lycaena (chalk-hill blue). Typical diurnal eye with strongly curved corneal facets and characteristic spaces between the pigment sheaths of the crystalline cones. Again there is a row of oval structures beneath the basement membrane as in certain moths and *Epinephele*. As in other butterflies, the retinulae are not strictly radially arranged but form graceful arcs at the edges of the eye. This arises because the lateral border of the eye is also curved (towards the interior of the eye). The retinulae follow this boundary so that they diverge distally rather like flowers in a curved vase.

2. Moths

Catocala nupta (red underwing, Fig. 15). The cones of this eye show indications of layering similar to that which I described in *Limulus*. Proximally there is a characteristic extension which is also cone-shaped. Were it not for the obvious separation, one could take it for a continuation of the rhabdom which possibly traverses the entire retinula. The nuclei reside in the thin part of the retinula. I counted six or seven; possibly there are always seven, as would be expected according to Grenacher. The rhabdom can be seen in the middle of cross-sections through the thick portion of the retinula (Fig. 15a). Near the basement membrane, nuclei and pigment lie irregularly arranged between the retinulae. The tracheae, though undoubtedly present, have been rendered invisible by the treatment with damar. Elongated oval structures (Fig. 15b), which stain well with picrocarmine lie behind the basement membrane (again the retinal elements of Thompson Lowne).

I did not record the state in which the animal (of Fig. 15) was killed. Most of the eye is quite unmistakably dark-adapted, but a group of ommatidia at the edge of the eye are equally unmistakably light-adapted. One of the latter ommatidia, with its iris pigment, is also represented on the right of the figure.

As in many other bi-functional eyes, the iris pigment can be divided into two components with different reactions to light. Thus, as is evident in the figure, a fine-grained pigment layer (*c*) remains fixed to the cone whatever the position of the real iris pigment. One could compare this layer with the iris tapetum in other eyes, although the only evidence of reflection is the dark brown colour around the black pseudopupil.

Leucoma (?) also has a typical moth's eye very similar to that of the red underwing (Fig. 15). The nuclei of the retinulae again lie very close to the tips of the cones, and the most distal layers of the optic ganglion contain similar, radially arranged, oval structures. The optical properties of the dioptric apparatus is shown in Fig. 57. The corneal facet (1), a section through the cone (2) and a cone on its side (3) are drawn as they appear under the microrefractometer.

Porthesia (Ocneria or *Cnophria*?).* A moth with a typical bi-functional eye, the retinulae with relatively long and thin distal portions. The crystalline cone appears to be bi-layered with some pigment lying at the boundary of the two layers (*a*).

That the iris pigment still projects beyond the tips of the cones when dark-adapted, is due, I believe, to my having checked for eyeshine with the ophthalmoscope shortly before killing the moth. At that time I was not aware that the pigment reacts so rapidly to light that this short illumination already causes a significant effect. Figure 31 shows the dark-adapted eye.

Lasiocampa quercifolia (Figs. 28, 29). The structure of this eye is in complete accordance with the moths already described. The migration of the iris pigment is again shown in the drawings.

Makroglossa (humming-bird hawkmoth; Figs. 17, 63) is a moth with a moth's eye, yet it flies in bright sunshine. There are long strands of pigment between the

* Due to a mislabelling of my preparations I am not sure to which of these moths the eye belongs.

retinulae which I have never seen in other moths. Unfortunately, I was not able to perform any experiments on the photomechanical action of the iris pigment, although I do not doubt that it behaves like other moths in this respect. Thus I sectioned eyes of two specimens and in one the pigment was unmistakably in the dark-adapted state, whilst the other was partially light-adapted. I have drawn ommatidia from both together in Fig. 17 (a, b). Unfortunately, the thin portions of the retinulae, as well as the muscle fibres of the iris pigment, were only preserved distally.

3. *Beetles*

The eye of *Cantharis fusca* (Fig. 9) is very similar to that of *Lampyris* (see also Chap. III). The crystalline cones are also fused with the cornea although in this beetle one sees a line of demarcation which stains strongly with acid magenta and picric acid. Within the proximal ends of the cones one sees small inclusions, again cone-shaped. Grenacher (1879, p. 130) considers these the real crystalline cones which are enclosed in a cuticular mass. Although this may well be true, functionally the entire structure operates just like the fused dioptrics in *Lampyris*, as can easily be verified in a brushed out eye. Those interested in the morphological significance of such cones may gain insight from another beetle, *Trichodes apiarius*. Here the proximal boundary of the cornea seems to merge into a laminated gelatinous mass – possibly a first step towards the situation in *Cantharis*. The firefly, where the enclosed cone is no longer visible, may represent a further evolutionary stage.

I wish to emphasize that it is not only the small enclosed cone which has significant refractive power, as one might infer from Grenacher's diagram (1879; Fig. 127), but that, as in *Lampyris*, the entire cone is optically effective.

Figure 9 shows cones with attached facets from a dark-adapted (*A*) and a light-adapted eye (*B*).

Colymbetes fuscus (water beetle; Fig. 16) has an eye very similar to that of *Dytiscus*. These are the only two insect eyes I know of where there is a clear effect of light upon the pigment, and yet the retinulae are of even thickness over their entire length, as in diurnal eyes. The nuclei lie distally just behind the cones. The corneal facets function as lens cylinders. These eyes thus form a counterpart to those of the rose chafer and its relatives, which have retinulae typical of nocturnal eyes but without pigment migration. Perhaps the explanation I offered for this anomaly in the rose chafer (p. 46) also applies to the carnivorous water beetles.

Dytiscus (great water beetle; Figs. 6, 7). Apart from its size the eye is so similar to that of *Colymbetes* that Fig. 16 could almost serve as a diagram for this beetle too.

Tropinota hirtella (Figs. 23, 24). The very small crystalline cones are difficult to see as they are embedded in pigment. They adhere closely to a thick cornea. The retinulae are thick proximally and thin distally, merging into the cone with a club-like extension containing nuclei. In cross-sections there is a reticular pattern between the thin parts of the retinulae reminiscent of that in butterflies and which probably also represents cross-sections of tracheae. An iris tapetum may be present in small amounts. Thus there are thin layers (Fig. 23 *T*) between the distal ends of the cones which reflect light more strongly than would be expected for a black pigment. Proximally the iris pigment ends abruptly, just as the retinal pigment does distally. I could see no change in either pigment layer under the influence of light or darkness (see p. 46 for more details).

Orycetes rhinoceros (for this animal I thank my brother Prof. Franz Exner, who caught and preserved it for me in India). The eye is very similar to that of *Tropinota*, though the cornea is thicker, the cones better formed and the individual elements of the eye generally larger. The position of the iris pigment and also the elegant cross-sections through the retinulae and the tracheae are similar.

4. Miscellaneous Insects

Libellula vulgata (Diplex) (Figs. 58, 59, 66, 67). The eye is divided into a yellow pigmented dorsal half, and a mainly black pigmented ventral part (Fig. 59). In the ventral eye half one can distinguish iris and retinal pigment although they are not sharply separated. Yellow cells representing the iris tapetum lie above the iris pigment. Just as the yellow pigment in the dorsal eye, these reflect light, though not so strongly as should be expected from a real tapetum. For further details see p. 77.

Both parts of the eyes represent typical diurnal eyes. The pseudopupils, which are discussed later (Chap. IX), are quite distinctive and allow one to distinguish the two eye halves immediately, though the transition does not appear so abruptly as in the sections.

The larva of a damselfly (*Agrion*) has an eye with structure and pigment distribution very similar to the ventral eye of an adult dragonfly (that in Fig. 59). It also shows similar optical phenomena with respect to the pseudopupils (Chap. IX).

Sections through the housefly (*Musca domestica*) eye were exceptionally elegant and clear. In particular one sees the position of different types of nuclei and the cross-sections of the retinulae and rhabdoms very well due to the lack of any opaque pigments in eyes preserved in alcohol. Staining was with picrocarmine. The configuration is the same as in *Eristalis* (Fig. 60) and thus I have no reason to give a special figure.

Eristalis. Strongly curved corneal facets and retinulae of constant thickness reaching up to the cones characterize the eye as that of a diurnal terrestrial animal. The yellow pigment (not tapetum) takes on a more cherry-red colour in the proximal part of the retinulae (not reproduced in the drawing). Black pigment is found only between the facets and in small hairs on the eye surface (*a* Fig. 60). Radial sections do not reveal the sausage-shaped tracheae separating the retinulae, but they can be well recognized in cross-sections (Fig. 60c), and best in squashed preparations. They form long, blind-ending sacks which arise from very thin tracheae. These thin tracheal branches penetrate the basement membrane and connect with the underlying tracheae (*b* Fig. 60).

The hornet (*Vespa crabro* Fig. 27) has a very thick cornea and small crystalline cones. As a true diurnal animal it has evenly thick retinulae in which rod formations can be seen. Along their entire length the retinulae are surrounded by pigment, which is particularly dense at both proximal and distal ends. Figure 27 shows three ommatidia from the lateral part of the eye where the optical axes are not perpendicular to the corneal surface. The figure shows the retinulae as curved. This is not an artefact of fixation or sectioning, but is the natural state arising from the configuration of the whole eye.

Locusta viridissima (the great green bush cricket). The eye is unmistakably diurnal and very similar to that of the hornet. However, the cones are ensheathed up to the tips in a strongly reflecting, yellow-green pigment (i.e. again a type of tapetum). The tips of the cones, as well as the whole retinula and the fenestration layer of the ganglion are richly pigmented.

Sirex gigas. The eye of this animal also strongly resembles that of the hornet. The individual cylinders of the very thick cornea and the cones are obliquely inclined to the corneal surface.

Tenthredo instabilis. As far as the individual ommatidia are concerned the eye is very similar to that of the hornet, though the cornea is thinner, and the cones are larger and perpendicular to the corneal surface.

B. Crustaceans

1. Macrurans[47]

Palaemon (Figs. 51, 52, 69, 70). As already illustrated by Schultze, a characteristic dome-shaped cap on the crystalline cone projects into a soft layer adjoining the back surface of the cornea. As in other macrurans, the cones themselves are made up of four pieces. Distally a short connecting element joins the cone to the cap. The cone mantle, lying proximally to the cone, tapers down to the retina and appears to be filled with a gelatinous fluid, coagulated into drops. However, directly adjoining the cone, this gel is of a somewhat firmer consistency and coagulates into a different and variable form. A coagulating, apparently rather fluid substance, is also found between the mantles. Whilst, for example in *Sicyonia*, the cones have spherical end surfaces, those in *Palaemon* are flat so that the dioptrics must rely largely on the action of a lens cylinder. In my preparations, the distal ends of the retinulae are not clear, though the nuclei at the same level are. Proximally the cross-section of the retinula is star-shaped and embedded in tapetal substance. Possibly there is a pigment layer within the tapetum. There is typical migration of both retinal and iris pigment. Notably, in this animal there are thread-like connections between the iris and retinal tapeta (Fig. 52). The latter is composed of the usual two layers. The corneal facets are square.

Crangon has an eye very similar to that of *Palaemon* with two tapetal layers in the retina and an iris tapetum. The retinulae have an elegant appearance arising from granular structures regularly distributed over their surfaces.

Nica edulis has an eye very similar to that of *Sicyonia sculpta*, and like this shows pigment migration. The only noticeable difference is that in *Nica* there is more tapetal substance distributed in fibres in the optic ganglion.

The bulldozer shrimp (*Scyllarus*) has well-formed, almost cylindrical crystalline cones. Proximally they have a nice lens-like curvature. The cornea is made up of square facets. The iris pigment surrounds the cones in thin, but rather compact layers and single thread-like (rather than sheath-like) processes project quite far proximally.

The material inside the cone mantle, and adjoining the convex surface of the cone stains differently to material outside the mantle, which gives the impression of a coagulated fluid in the preparations. When light-adapted, the retinulae were more elongated with spindle-shaped swellings, in the dark-adapted preparations they approximated a cylindrical shape. I consider the former shape the natural one, and the latter (which I also saw in one light-adapted eye) the result of poor fixation. Cross-sections through the spindles (light-adapted eye) show the rhabdom as a seven-armed star surrounded by pigmented retinula cells.

I could confidently see migration of the retinal pigment, but not of the iris pigment. In the dark, the retinal pigment is found only around the proximal parts of the retinulae – the distal parts remain free. In the light the pigment also covers the distal parts and shows a characteristic streaky appearance, the explanation for which is found in the star-shaped cross-section of the retinula. The layers of the optic ganglion were entirely devoid of pigment.

Sicyonia sculpta (Figs. 53, 54, 39) has a very nice, easily-sectioned eye. There are three tapetal layers and clear migration of both iris and retinal pigment. There is a large space between the cones and the striated rhabdoms.[48] This space is traversed by the cone mantles which, in the sections, are separated by a coagulated fluid. The cones themselves have nicely rounded front and back surfaces. The facets are square.

In the light the iris pigment and iris tapetum undergo extensive proximal migration whilst the retinal pigment from the optic ganglion migrates through the basement membrane and the tapetal layer, ensheathing the rhabdoms up to their distal ends.

The large spherical eyes of the spiny lobster *Palinurus* (Fig. 50) have a clear vitreous zone which can be dissected with the naked eye. As in vertebrates the retina lies behind it. The retinulae are rather small. The structure of the eye is very similar to that of *Palaemon* or *Sicyonia* except that even in the dark the retinal pigment remains in the retina, although behind the tapetum. In the twilight-adapted eye all the pigment lies distal to the tapetum. The left hand ommatidium in Fig. 50 represents the stage where the pigment is just in the process of passing through the tapetum. It has not yet reached the distal parts of the retinulae but some of it already lies between these and the tapetum. The lateral part of the same eye shows the retinal pigment still in the dark-adapted state (Fig. 50, right). Although the iris pigment was not seen to migrate, I have no observations from a completely light-adapted eye.

The contents of the cone mantles (seen in cross-section in Fig. 50b) appear strongly refractile and homogeneous near the cones, but more proximally they break up into a drop-like coagulate. No facets can be recognized on the front surface of the cornea; on the back surface there is a soft hypodermis showing clear square facets.

Although my preparations of *Astacus fluviatilis* were not good enough to warrant a figure, the iris pigment was clearly seen to migrate proximally in the usual manner on light adaptation. Similarly, the retinal pigment moves to surround the distal ends of the rhabdoms, whilst in the dark it has completely vanished from the retina and is concentrated in copious masses behind the basement membrane. Even in the light a proportion remains here. A typical tapetum envelops the proximal halves of the retinulae in both the light and the dark.

Peneus membranaceus (Figs. 47, 75). The splendid eyes of this animal are exceptional in having a fine tapetum but virtually no pigment. In one dark-adapted eye and one eye fixed in diffuse daylight (both fixed in alcohol) there was a very nice iris tapetum but no trace of iris pigment. However, I did find traces of a brown pigment in one light-adapted eye fixed in Müller's solution. It occupied the usual position of the iris pigment stretching from the iris tapetum to about halfway along the cone. This pigment layer was very inconspicuous and I could not decide whether this individual had iris pigment and the others not, or (as seems more likely), whether all specimens possess some pigment interspersed with, and hidden by the iris tapetum, and that the pigment is only sufficiently separated from the tapetum to be confidently recognized when light-adapted. Finally one must consider that the differences between the eyes were due to the fixatives, but it is far more likely that Müller's solution would dissolve out the pigment than the alcohol.

Insight into these questions came from specimens of the same species from the Imperial Natural History Museum kindly shown to me by Herr Kölbel. Although they had lain many years in alcohol, these animals (caught on the Spanish coast) showed clearly pigmented eyes. There thus appears to be no doubt that there are individual variations, possibly correlated with the depth at which the animals live.

Proximally, the short striated rhabdoms invade the retinal tapetum. As in many crustaceans, *Peneus* has a second, completely separate (though rather sparse) tapetal layer in the optic ganglion. It is striking that this second layer was not seen in the light-adapted eye preserved in Müller's solution.

The cones are proximally rounded. Each cone process ends in two strands which are filled with a coagulated fluid. The corneal facets are square.

2. *Anomurans*

The hermit crab (*Pagurus*, Figs. 71, 72) has hexagonal facets. The migration of the iris pigment can be demonstrated in microscopic preparations – as is to be expected from the behaviour of the eyeshine.

3. *Brachyurans*

Dromia vulgaris (Figs. 55, 56) has thick square facets. As is often the case, a lens-shaped refractile body can be seen within the crystalline cone. I do not know if this has the same optical properties in vivo. The distal extensions of the retinulae adjoin the cones. Migration of both iris and retinal pigment is typical.

Portunus (Figs. 37, 38). The cones are heavily coagulated, so that, although I have drawn their shape as the majority appear, I cannot vouch for accuracy. The facets are hexagonal. The behaviour of the iris pigment is noteworthy. In the dark it has a wide aperture leaving the proximal curved ends of the cones free, and in the light it moves over the retinulae, adhering tightly to them like a crust. Usually the pigment forms a tube as wide as the cone, i.e. far wider than the retinula or cone mantle which lie within.

The behaviour of the retinal pigment is also very striking. It migrates from the optic ganglion and the proximal halves of the spindles to the distal halves of the spindles, leaving only sparse remains in its original position. One can directly see the strands containing the pigment. They surround the rhabdoms and penetrate the basement membrane disappearing into the optic ganglion. In cross-sections (*a,b*) these strands form stars, showing that they are ribbon-shaped with their broad sides arranged radially.

Galathea (Figs. 45, 46, 65). The cone mantles form long tubes between the cornea and the cones proper, and are filled with a coagulated fluid. Both iris and retinal pigment migrate. In the dark-adapted eye the spindles of the retinulae are contracted (probably as a result of fixation), so that the tapetum does not rest on the basement membrane. The facets are square.

Maja (Figs. 41, 42) has retinulae extending almost up to the cones. There is no tapetum and migration of the iris and retinal pigment is minimal. All one can see is that the round nuclei in the distal retinula become surrounded by iris pigment in the light, and some migration of retinal pigment out of the optic ganglion is very probable but not certain. The facets are hexagonal and the cornea shows a marked lamination. The cones were poorly preserved in my preparations.

Pisa (Figs. 48, 49). The laminated cornea has hexagonal facets; the cone tips are surrounded by tightly adjoining nuclei. The retinula is divided into two parts; as usual, the nuclei are contained in a distal swelling. Tapetal substance is found between the proximal ends of the retinulae and also in the optic ganglion. Cross-sections (Fig. 48, *d*) show that it is in all likelihood stored in cells with nice round nuclei. There is clear migration of both iris and retinal pigment. In the dark the latter lies almost exclusively behind the basement membrane, and in the light around the distal halves of the spindles where it is arranged in elegant continuous threads. In one light-adapted eye I also found pigment around the proximal halves of the spindles where cross-sections reveal beautiful pigmented stars with a central transparent rhabdom. The arms of the star were formed by the pigment threads. In both thick and thin parts of the retinula the rhabdom appears to be composed of four discs. Remarkably, the distal end of the retinula is separated from the cone in the light-adapted eye, yet still appears attached to it via a thread. Both light-adapted eyes showed this behaviour, although one had been fixed in alcohol and the other in Müller's solution, so that a fixation or shrinkage artefact seems unlikely.

At least in my preparations the tapetum reflects light poorly, so that it appears less bright in incident light and less dark in transmitted light than in other animals.

4. Miscellaneous Crustaceans

Squilla mantis (Fig. 22). I was unable to observe either a tapetum, or any light-induced pigment migration. The facets are hexagonal.

The eyes of *Squilla*, and also those of *Phronima* and *Copilia* have such an unusual design that they are discussed in a separate chapter.

Chapter VIII The Eyes of *Squilla, Phronima and Copilia*

So far, the eyes of insects and decapods have been almost the exclusive subject of discussion. Apart from *Limulus* I have very little experience of other crustaceans; however, judging from the descriptions I read, some of them seemed so interesting that I included them in my study.

A. Squilla mantis[49]

The individual ommatidia of *Squilla mantis* are very similar to those of brachyurans and anomurans. I succeeded in obtaining fairly good images of the ray path in the fresh eye. It is very similar to that in *Limulus*, i.e. one sees a somewhat vague inverted image of external objects in the vicinity of the cone tips. With two point sources one sees that the rays from each cross in the image plane with a very fine angle, and that the cones of light proceed further, approximately parallel to each other, without crossing (at least for the distances relevant here). This behaviour is the same for ommatidia in both the most highly, and the most weakly curved parts of the eye. The animal thus sees with an apposition image.

What makes this eye exceptional is its shape. It resembles a cylinder with rounded ends and a ring-shaped constriction in the middle (Fig. 22).

Incidentally, an eye with a similar shape is also found amongst the insects, namely *Ascalaphus* (Prof. Brauer pointed this out to me after I had finished this study).

In general, a long-section through the *Squilla* eye is kidney-shaped, and a cross-section circular, as long as the eye stalk is not included. No section results in a straight eye margin since the curvature at the end of the cylinder continues directly into the curvature of the constriction. This is important since it is a principle of the compound eye that the ommatidia are inclined with respect to each other.

Although I have already alluded to the distortion of the upright retinal image necessarily caused by the shape of the eye, it is worth considering this and its possible significance in more detail.

Let us first consider only half of the eye, reaching from the end of the cylinder to the constriction. If the cylinder is horizontally aligned, then the radius of curvature is much greater in the horizontal than in the vertical. The retinal image of a square must then have the shape of an elongated rectangle with its long side horizontal. Of what use could this be to the animal?

We often exploit such an image distortion in special instruments. Thus thermometers are now commonly ground so as to magnify the thread of mercury in one dimension only. The thread thus appears thick and is easier to read. Magnification in the other dimension would be superfluous. By the same principle a bright line is used to produce a spectrum. The spectrum from a point source would be just as long and have the same order of colours, but many details would be missed without the vertical magnification.

In the same way, the retinal image in *Squilla* will still resolve a vertically arranged system of lines which could no longer be resolved when horizontal. The peculiar shape of the eye can thus be explained if it is assumed that it is important for the animal to distinguish details arranged in a certain orientation.

The other peculiarity, the ring-shaped groove dividing the eye into two similar halves, must result in an object projecting two images in the same eye. This is inevitable since, apart from trivial exceptions, any theory of compound eye function requires that a point projects an image in that ommatidium whose axis is aligned with the point. Here there are at least two ommatidia on either side of the groove with axes directed at the same point in space. The animal thus sees two images with one eye. The most important advantage of binocular vision is undoubtedly the possibility of judging distances and *Squilla* thus has this potential with each eye alone. Of course, we do not know how well other insects and crustaceans can perceive depth monocularly, but it is clear that *Squilla* has at its disposal one of the most important factors lacking in the others.

It should be noted that, just as with our binocular vision, the constriction of the *Squilla* eye would also be useless with respect to a horizontal line. This is also the reason for the well-known tendency of birds to fly into telegraph wires.

For simplicity we have assumed above that the eye has its long axis horizontal, yet it is clear that whatever the position of the eye, the object orientation projecting the clearest retinal image also results in the best estimation of distance. One could hardly call this coincidence accidental.

In conclusion, the peculiarities of its eye enable *Squilla* to see structures in which a particular orientation predominates more clearly and to judge their distances more accurately than would otherwise be possible. I know too little about these animals' life-style to even guess what these structures might be.

B. *Phronima*[50]

It has long been known that the eyes of the Phronimidae differ greatly from those of other crustaceans. Again, each eye is divided into two parts, which are so different that one could justly consider each as a separate eye (Text Fig. 20). There are thus two lateral eyes with conventionally located visual fields of normal extent, and two medial eyes with exclusively dorsally directed visual fields. Correspondingly, there are also four retinae (*R1, R2*). Text Figure 20 shows the right half of a head seen from the front at low magnification. Since these animals are almost transparent, one immediately recognizes the lateral eye with its short crystalline cones and retina (*R1*), and also the enlarged medial eye with cones projecting long threads to the second retina, *R2*. The retinulae almost appear to be extensions of the cones and together resemble a cornet.

Here again I must thank Prof Brauer for the information that a similar division of the eyes is found amongst the insects (namely in the ephemerid *Chloe binoculata*).

With respect to the structure of the eye of *Phronima* I refer to the work of Claus (1879), from which Text Fig. 20 is adapted, and discuss here only those features which I believe to be of physiological significance.

Whilst the proximal ends of the cones in *Gammarus roeselli* lie at some distance from the retina so that one cannot rule out a superposition image, the cones in the otherwise similarly built lateral eye of *Phronima* project so far into the retina that their very small, flat end surfaces lie right up against the distal ends of the rhabdoms. Since these have almost the same thickness they look like extensions of the cones (Fig. 43). The central processes of the retinal elements penetrate the basement membrane (*membrana fenestrata m.f.*) and contact the cells of a primitive optic ganglion. The optic nerve is composed of nerve fibres leaving these cells. The eye contains no pigment apart from a diffuse yellow coloration of the retina.*

I know of no other animal where the dioptric apparatus, the retina and the neural elements are so simply and clearly laid out.

The optical processes involved are less clear. There can be no superposition image for reasons already mentioned, but a normal apposition image must also be ruled out. This requires that an, albeit poor, dioptric image of an external object be projected near the tip of the cone. However, first the cones are not of regular enough construction – as Schmidt (1878) pointed out they are far from being circularly symmetrical; secondly, at least for the medial eye, the tremendous length and narrow diameter of the cones completely excludes a normal dioptric image. The cone merges gradually into a shank 15 times longer than the swollen distal end of the cone. (Fig. 44 shows an isolated cone with only part of the thread still attached.) For a dioptric image to arise, the thread should not deviate more than about one thousandth of a millimetre from a straight line over its entire length. This is certainly not the case; one sees instead asymmetries and bends which certainly also occur in real life. From this Schmidt deduced that the cones in *Phronima* guide light by total internal reflection like a glass rod, and that the mosaic theory of vision does not apply here. As already pointed out, the latter deduction is false; in a certain sense the former is correct.

In my first paper on the compound eye (1875) I proposed the theory that each ommatidium guided axial rays by a combination of refraction and reflection to the tip of the cone where they were then absorbed by the retinal elements. By and large this theory was wrong. At that time I did not know of the optical effects of lens cylinders. Now, however, it must be upheld for *Phronima*, although in a form improved by the knowledge of lens cylinders. The passage of rays in glass rods also occurs when they are curved, indeed this principle has recently been put into practical use.

* As seen in animals preserved in alcohol,sometimes prefixed in osmic acid. I had no opportunity to study fresh specimens.

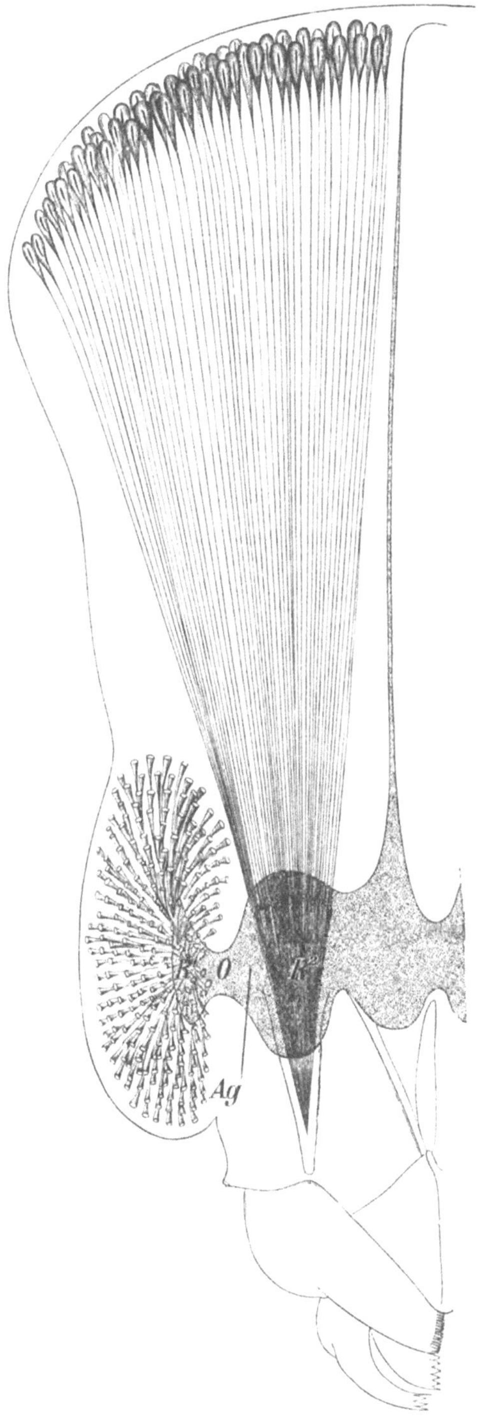

Text Fig. 20

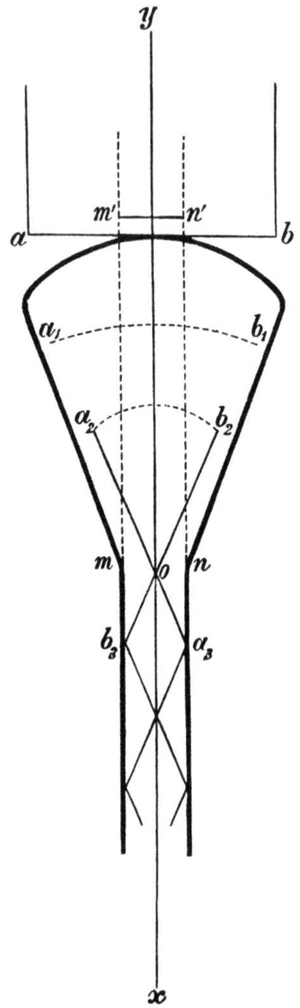

Text Fig. 21

Let us first consider the radial array of the distal ends of the crystalline cones and let us assume that rays incident approximately along the axis of such a cone reach the thread through which they are guided to the retinal element. It is only necessary that the retinal elements have the same respective arrangement as their associated cones and an upright image must arise whether the cone threads are straight or bent. Now it can be convincingly seen in the medial eye that the cone threads converge perfectly regularly onto the retina without crossing or interweaving so that this corresponding arrangement of cones and retinal elements is in fact guaranteed.

That axial light is really guided to the cone tip could be shown in the lateral eye as follows. The eye was stained, embedded in celloidin and cut into thick sections. A perpendicularly oriented cone was selected and its distal end directed at the microscope mirror. Just as with a glass rod the proximal end now appears bright or dark as the distal end is illuminated or shaded. The light itself proved that it had travelled the length of the cone as it was coloured deep red, whilst other cones, which were not perpendicular, were barely coloured at all.

As happens to every investigator, I also found that things were more complicated than I had imagined. Namely these cones which, as Claus correctly describes, consist of two adjoining pieces, are built on the principle of the lens cylinder, and the convergent effect of this will reinforce the function of the outer convex cone surface.

I showed above (p. 40) that a geometric cone is not a suitable structure for guiding light to its tip. However, here we are dealing with a structure specifically designed to solve the problem of guiding axial light into the cone thread. Text Fig. 21 shows the situation somewhat schematically. Without refraction the rays entering the thread (*mn*) would be practically restricted to the ray bundle *m'n'*. Probably some neighbouring rays would also enter through reflection, but for such a thin thread these would be very few. However, with refraction the entire bundle with cross-section *ab* corresponding to the width of the cone can reach the thread. Thus the plane wave surface *ab* will be curved into a_1b_1 by refraction at the convex surface, and this curvature will be further increased to a_2b_2 by the lens cylinder. Since rays are always perpendicular to the wave surface, a ray cone a_2o, b_2o will reach the thread and be guided by internal reflection to the retinal elements. This ray cone contains all the light which enters the convex surface of the cone axially. The cone-shaped distal extension, with its convex surface and lens cylinder construction, thus achieves a very considerable increase in the brightness of the retinal image.

Rays entering the cone obliquely will either be eliminated or may also reach the cone thread if their inclination to the axis is sufficiently small. Thus some light immediately surrounding the axis will also reach the thread.

In conclusion, *Phronima* also sees with an upright retinal image. The image is related to the apposition image but differs in that reflection as well as refraction is used to guide the the light in each ommatidium. Dorsally the animal sees with a more flattened eye, probably analogous to the situation in dragonflies. At any rate, the eye as a whole is much less perfect than that in other crustaceans.

C. *Copilia*[51]

Up to now, despite their peculiarities, all eyes discussed did at least have an upright retinal image. However, the recently described eye of *Copilia* and its closest relatives (especially *Sapphirina*) is really most extraordinary and appears to fit into no known classification.

Copilia, which I had the opportunity to study dead and alive, is a dorso-ventrally flattened copepod, a few mm long and shaped approximately like an isosceles triangle. At both edges of the anterior end of the animal (the base of the triangle) there is a conspicuously beautiful lens. The distal convex surface faces

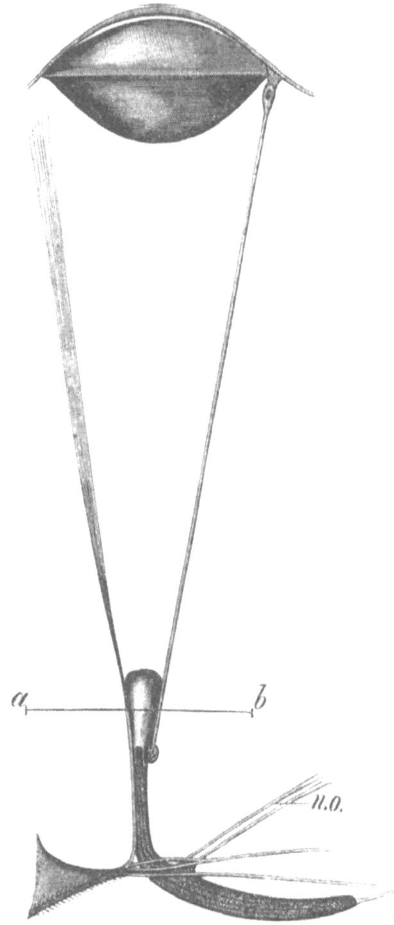

Text Fig. 22

water and the posterior surface faces the interior of the almost completely transparent body (Text Fig. 22).

As observed by Grenacher, the lens has two components: a cuticular concave-convex lens, and a stronger biconvex lens lying behind. The lenses form the most anterior part of the whole animal, and behind them there is not, as one might expect, a retina, but only transparent tissue. One first notices a structure much further back – about half the length of the body – which at first sight appears to have nothing to do with the lens. It is similar to a crystalline cone, distally curved, proximally blunt and of high refractive index. It rests on a knee-shaped, bent yellow rod which is the sole pigmented part of the animal. Closer observation reveals that this "cone" is attached distally by strands reaching almost as far as the lens, also that a nerve, the optic nerve (*n.o.*), leads into the side of the yellow rod and that a striated muscle is attached to it. Grenacher, to whose anatomical description I refer (Gegenbauer, Leuckart, Claus and Haeckel also studied this eye), further observed that the yellow rod (here drawn in grey) contains the analogues of the retinula cells and the rhabdoms – probably three in number.

It is easy to understand the interest excited by such an eye. It has been compared to a telescope, the objective being the large lens and the ocular the refracting body, which for simplicity I have called the crystalline cone. This may well be, but where is the retina which should receive the telescopic image? The image from each lens must have a considerable size, and only the smallest part could be received by the cone, and even for this the necessary number of nerve endings are not present. With reference to the performance of the eye, Grenacher (1879, p. 73) writes: "... there are a variety of reasons for considering the eye as analogous to a single element of the compound eye. Of course with logical application of present concepts, particularly with respect to the performance of the rods and the significance of their number for visual resolution, the performance will be reduced to a minimum. The vestigial perceptual apparatus, the anatomy of which excludes image perception, contrasts strangely with the comparatively colossal expense involved in producing the image, particularly evident in the construction of the lenses".

I also agree that such an interpretation of the function of the eye cannot be considered satisfactory. What, with our present knowledge, could be the use of a single ommatidial element? The analogy with the structure is also questionable; only the pigmented rod with its retinula cells and rhabdoms bears the similarity to the retinula of an ommatidium emphasized by Grenacher. I believe I can give a both simple and satisfactory solution to this riddle.

From pictures and descriptions of this eye I had already come to the following conclusion: the lens must, at any rate, project an inverted image of external objects. This image could be sampled by a retina which was simultaneously excited by different parts of the image, but such a retina is lacking. There is only one retinal element joined to the cone; however, this could be successively stimulated by different parts of the image if it is scanned along it. The single retinula would then, for example with respect to a horizontal row of points, see alternately bright and dark, and thus be in a position to determine their order. In a given position it would detect light and this position would characterize the location of a bright space just as well as the location of an excited retinal element in a vertebrate or compound eye. Indeed, with its single retinal element, the animal would be doing just what we do with our fovea, which we also traverse along the image by means of eye movements, inspecting interesting parts more closely. Of course, we also have peripheral vision which is lacking in *Copilia*.

If this theory of the function of this eye is correct, two conditions must be met: first the retinal element (the yellow rod with attached cone) must lie in the focal plane of the lens, and second, it must be able to move within this plane, i.e. it must remain at approximately the same distance from the lens.

In the Naples zoological station I had the opportunity to study *Copilia* alive and the first look down the microscope convinced me that the second condition was fulfilled. The bent rods showed the most lively movements, which became positively convulsive when one covered the well slide with a coverslip, whereby the animal probably began to experience oxygen deprivation. However, the movements were always noticeably regular and symmetrical. Both rods were pulled to and from the median plane, and always stayed – as far as can be judged without measurement – equidistant from the lens. The movement is reminiscent of a man performing an exercise with dumb-bells, except that the rods remain bent and only change their position within the median plane.

The excursion of the rod was approximately one quarter of the animal's width. Whether there are vertical movements as well as lateral is not certain, but because of the minimal depth of the animal's flattened body they could not be very extensive.

The movements are undoubtedly achieved by the striated muscles described and illustrated by Grenacher; the rapidity of the movements further confirms the involvement of striated muscles. I do not know if certain strands attached to the rod are also muscles, as has been claimed. Thus, they could also function as elastic bands affecting the rebound of the rods towards the midline when the muscle relaxes. The arrangement of these strands – one attached to the finger-shaped end of the rod and the other clasping the rod near its attachment to the cone would speak in favour of this.

The other condition, that the retinal element must lie approximately in the focal plane of the lens, can be checked by measurement. With a micrometer I determined the distance between the rear surface of the lens and the front surface of the cone as 0.87 mm in the living animal.

It is less simple to measure the distance of the image from the lens. I cut off a small strip from the animal parallel to the anterior surface; with some patience I succeeded in placing this on its edge in water so that the back surface of the lens was directed towards the microscope objective. One now sees a surprisingly clear and rather large image. Its distance from the back surface of the lens, measured with a micrometer screw, was 0.93 mm. The difference between the two measurements, 0.06 mm, is insignificant compared to the total distance of almost 1 mm, and there can be no doubt that the cone lies almost exactly in the focal plane of the lens.

We now have to consider the function of the refractile cone and its distal convex surface. When discussing the *Phronima* eye I described the value of cone-shaped structures for the collection of rays approximately parallel to the axis, and the great significance of the cone for the exclusion of oblique rays has been expounded in Chapter II (p. 40). Due to the large distance of the lens from the image (in relation to its size) rays intersecting on one image point will be virtually parallel. If they strike the convexity of the cone they will most likely emerge together at the apical surface thus reaching the retinula even if the focus is slightly before or after the cone's surface. This means*, that no rays can reach the retinula except those coming from the vicinity of the lens, i.e. those important for vision. Rays striking the cone with a small inclination (i.e. those coming from near the lens) must be refracted to the proximal cone margin where they will either be absorbed by the characteristic pigment layer there**, or possibly actually reflected, thus blurring the clarity of the image. However, more steeply inclined rays will be so refracted that they exit from the cone without reaching the retinula .

* Because of the reversibility of ray paths.

** However, the pigment would have to adhere to the cone material like, for example, the black lacquer on the glass of black mirrors.

The function of the overlying cone is immediately clear when one considers that the animal is completely transparent apart from the retinula. Thus, even though the distal end of the retinula adjoins a short pigmented tube, light from a large part of the visual field could still reach it. However, the cone ensures that a large proportion of the ray cone from the lens is guided to the retinula whilst virtually all light from other directions is excluded.

Copilia thus sees by sampling the image projected by the lens with one retinal element. The psychic process underlying the recognition of objects is basically the same as that which we use to recognize forms by running a finger along an object and reconstructing the Gestalt from the sequence of sensations. Such vision has a certain analogy to our perception through eye movements.

Because of its body shape the animal's visual field (i.e. the area it can sample) is elongated. At a distance of ca. 2 metres it should be able to see along a line 1 metre long (successively of course).*

The retinal image is inverted, and since the movements of the two retinulae are mirror-symmetrical, the animal sees separately with each eye and thus does not possess binocular vision, in contrast to virtually all other crustaceans.

* This is naturally only an approximation; I could be in error by 50%. Nevertheless, it gives an idea of the order of magnitude.

Chapter IX Accessory Optical Phenomena in Compound Eyes

Up to now the optical processes which I have discussed have concerned the function of the eye as a visual organ, in particular the projection of the retinal image. However, it is clear that such a complex optical apparatus must also give rise to reflections, refractions, absorptions, etc., which have nothing directly to do with vision, but which, because of the regular structure of the eye, can cause regular optical phenomena – like the Purkinje images in human eyes (deriving from reflection) or the eyeshine of cats. The diversity in structure amongst compound eyes means that these phenomena are different in almost every species; nevertheless, certain types occur repeatedly. These will be discussed here and the often complex explanation given. Isolated observations already exist, but to my knowledge no general explanation of these phenomena has been attempted.

In particular I will discuss the phenomena of eyeshine and pseudopupils.

A. Eyeshine

Leydig (1864a, p. 255), who discovered the tracheal tapetum, supposed it to be particularly well developed in the convolvulus hawk moth (*Sphinx convolvuli*) since: "Lepidopterologists (Kleemann) report that its eyes shine like glowing coals in the dark." Only later did he succeed in procuring this animal alive, and he then describes the following:

"As I caught the first living animal in twilight and directly examined the eye I was surprised to see no hint of any glow. The eyes had a dark appearance similar to that of beetles. After arriving home I again examined the moth both in bright light and complete darkness but still the eyes did not glow. Tired of the experiment I left the animal in an empty drawer. The next day at the crack of dawn, I carefully opened the drawer; the moth was resting quietly at the edge, I looked at its eyes – and they now shone beautifully 'like glowing coals'. Thus I believed to know the conditions under which the glow appears. The moth's cage faced away from the light which came through the one open shutter, and the eyes only glowed when a little light fell on the animal through the crack of the open drawer. As is well known one also sees the eyeshine of, for example dogs and cats, under similar conditions. Curiously however, a few hours later, although I used exactly the same conditions, there was not the slightest trace of any glow. The eyes again had the dark velvety appearance of the previous evening. I observed the same variability with other freshly caught animals."

From such observations Leydig correctly concludes that, apart from the appropriate illumination, the eye must have a certain internal state for the glow to appear. He considers two possibilities: either an expansion of the tracheae forming the tapetum, or "a change in the state of of the pupil layer". By pupil Leydig refers to the narrow openings in the pigment through which the tips of the cones or the distal ends of the retinulae project, which (according to his observations) are able to contract. We now know that this view, at least for moths, is not tenable (see Chap. III), but we will see that Leydig's second alternative is in principle correct.

Whilst Leydig does not give further details about the occurrence of eyeshine in lepidopterans, and merely wonders that he could find no glow in certain species which also have tracheae in the eye, Schultze makes the statement that the eyes of moths glow in the dark and those of butterflies do not. With the reservation that only a fraction of the many thousand species have been investigated, we will see that this statement, at least under certain conditions, is correct. Incredibly, though, Schultze refutes Leydig's easily reproducible observation that the eyeshine is only short-lived. He states that the eyes always glow in the dark and attributes Leydig's finding to inappropriate illumination. Further, he attributes the glow, not to the tracheal layer, but to the laminated structure of the retinulae which he himself described.[52]

Kühne (1877) finally proved that the eyeshine disappears during strong illumination and reappears after a long period in the dark. Thus he directly observed the glow fading away whilst illuminating a death's head hawkmoth (*Acherontia atropos*) with an ophthalmoscope. From the sluggishness of the change he supposes that it is related to processes in the pigment cells since it was too slow for muscles and too fast for a change in the tracheae.

1. My Own Observations: Insects

When studied with the ophthalmoscope, not all, but certainly the majority of compound eyes show the ability to glow, i.e. to reflect light entering the eye according to certain principles in full analogy with the so-called eyeshine of vertebrate eyes. In vertebrates the glow emerges from the pupil, and similarly, in the compound eye the glow is restricted to a circular region more or less coincident with the so-called pseudopupil described later. I therefore refer to the phenomenon as the luminous pseudopupil.

A basic difference from the vertebrate eye is immediately seen when the compound eye is rotated whilst observer and direction of illumination remain constant. The position of the glow with respect to the observer does not change, meaning that during rotation the group of facets which glows changes. When the eye is optimally illuminated the facet whose optic axis coincides with the direction of observation – and a circle of surrounding facets – appear bright. The size of the bright circle depends upon the fine structure of the eye and the position of the two pigment layers. It can, as in dark-adapted moths, be several millimetres across (the cases described by Leydig, Kühne and others); it can also, as in butterflies, be so small that it can only be seen under optimal conditions, and then usually only with the aid of magnification. My failure to detect a glowing circle in some insects and in most light-adapted crustaceans was probably at least sometimes due to the size being beyond the limit of resolution. I presume this because, with some insects, after first searching in vain, I did in fact eventually find a small bright point. The colour of the bright circle is normally red but in certain species can also be whitish or greenish and can change during illumination. Although I have not illustrated the eyeshine in insects, Figs. 68 and 70 (for crustaceans) are very similar to the narrow and broad eyeshine of butterflies and moths respectively.

The optical processes in vertebrate eyes are basically similar whether there is a tapetum or not. Nevertheless, because of the more conspicuous optical phenomena one usually refers to those eyes with tapeta as "luminous eyes". This is also the case with compound eyes; like the human eye the fly's eye glows under suitable conditions. In both cases though, the glow is insignificant compared to that from animals with a tapetum – on the one hand carnivores, certain ruminants etc., and on the other, moths and nocturnal crustaceans.

I start with the moths, these being animals where the phenomenon is already known and most obvious. Despite the great diversity though, the explanation is essentially the same amongst different classes.

When one captures a moth after dark and holds it behind a flame so that the observer's eye, flame and moth's head lie almost in a straight line (i.e. the head is seen just at the edge of the flame) the eye shines. The glow has justly been compared to that of glowing coals. The phenomenon is seen even more beautifully and more practically with the use of an ophthalmoscope (I usually equip mine with a magnifying glass). One then easily recognizes that only a circular portion of the eye glows, and that on turning the animal, this disc moves across the eye so that it is always directed towards the observer. When moved to the edge of the eye it disappears gradually. If one observes the animal head-on, a luminous pseudopupil is seen in each eye; however, because of the marginal position it is now lens-shaped.

The explanation for the glow is fully analogous to that in vertebrates. As we have seen, the tapetum between and behind the retinulae reflects light which has passed through the retinulae. On the way back out the light will pass mainly through the retinulae, or rhabdoms, for a second time, thus increasing the sensitivity of the eye at low light levels. However, from the structure of the tapetum some of the light must also be scattered diffusely. In dim light (i.e. the normal situation in the dark-adapted eye), this will be negligible, but in bright light (e.g. with a flame) this diffuse light must become noticeable.

Let us recall the normal ray path in a dark-adapted insect eye with a point source as object (Text Fig. 11 p. 27). The iris pigment lies between the cones and the rays of a large cylinder of light will be refracted through the dioptrics to meet at the image point *B*. Since this lies in the strongly reflecting retinal layer, it will also serve as a light source and rays from it reaching the observer's eye through the cones will be refracted exactly as the incident rays were. Thus the path of the rays emerging from the illuminated retina are shown by the lines $x'B',p\ Bq$ (Text Fig. 11) and all the lines lying between them. The result is again a cylinder of rays returning the same way as the rays entering the eye from the object. If the situation was as mathematically precise as just described, one would not be able to see the glow with the naked eye, but only with the aid of the ophthalmoscope, since the observer's eye must coincide with the object. However, this ideal situation is not realized in compound eyes and it thus suffices when the observer brings his eye only approximately in line with the light source and the compound eye. On the other hand, this behaviour explains why the glow is not seen at all with inappropriate illumination and is best seen with an ophthalmoscope which allows alignment of the light source (or its mirror image), observer's eye and specimen.

In the ideal situation the eyeshine should disappear as soon as the observer moves his eye from the position of the point source, but as we saw, this happens only after a considerable misalignment. If one observes the eye with the ophthalmoscope from different angles whilst fixating on a point, e.g. the middle of the eye, one can see that, even when the light source is a point source (e.g. the reflection of the sun in a Helmholtz ophthalmoscope) this point still appears bright whilst one describes an arc of several degrees around the eye. Since the luminous pseudopupil is rather large, the position in which the middle of the eye first moves into the pseudopupil is quite different from that in which it moves out. I have not measured this angle since it depends upon the species and the state of the pigments in the eye.

This observation shows that even with a point source, the rays leaving one facet are not parallel, as predicted in the ideal situation, but that they form a cone of light. This enables us to see the eyeshine without the aid of an ophthalmoscope, but how is this ray cone formed? In the vertebrate eye a ray cone is only formed when the retinal image is of considerable extent.

The question is easy to answer. It could be that under these conditions, when a dazzling image is projected into the eye with the pigment in the fully dark-adapted state, it simply loses its point nature by scattering. However, another factor is probably more important. Compared to the vertebrate eye the retina in the compound eye is very thick and the tapetum is also distributed over different levels. At whatever level the perception of the retinal image may occur (if it occurs at any one level at all) there is no doubt that both the retina, and especially the tapetum, will be illuminated at very different levels in the eye, so that the returning rays will impinge upon the dioptric apparatus from different distances. If the point *B* (Text Fig. 11) represents the most distal end of the retinula, not only light from here must return to the crystalline cones, but also light from more proximal regions, e.g. from the various levels of the tracheae which surround the proximal ends of the retinulae. It is clear that rays from different depths must reach the axis of the cone (e.g. point *n*) with different angles and must thus emerge from the cornea with different angles. This is apparently the main reason why a complete ray cone leaves the eye from each facet. All else being equal, the angular size of this ray cone determines directly the size of the luminous pseudopupil.[53]

Thus the phenomenon of eyeshine can be explained along the same principles as in vertebrate eyes, but differs in that instead of a pupil there is a pseudopupil which changes its position with the observer, and the bright objects within the eye lie, not in one, but in many different planes so that the rays leaving one point of the eye diverge in a ray cone even when the light source is punctate.

The fading of the eyeshine during illumination is a general phenomenon in nocturnal animals. Of the many moths I have investigated, when caught in the evening all showed eyeshine which disappeared when they were brought into a bright room. The speed with which this happens probably depends on several factors, in particular the degree of illumination. A noctuid which flew into my room in the evening showed no more eyeshine after 25 minutes, at least in that part of the eye which I had repeatedly investigated with the ophthalmoscope. The change in the opposite direction takes longer. In my experience it always takes at least an hour before a non-glowing eye glows again. Doubtless the speed of transition also depends on the animal's health. Feeble, dying moths lose the ability to regain their eyeshine. In contrast the transition to the non-glowing state appears to function reasonably in very sick specimens, as the following experiment with the lappet (*Lasiocampa quercifolia*) shows (see p. 47). Half of the head of a dark-adapted specimen is cut off and brought into the sun. The remaining eye, which at first glows beautifully, became completely dark within ca. 15 minutes.

It should be noted that to achieve eyeshine in the day one must be very careful to exclude light, and a small crack in the cage is sufficient to prevent the transition to the dark-adapted state.

Kühne has correctly described the manner in which the eyeshine fades: the glowing circle becomes progressively smaller and somewhat paler. It also shows some irregularities in the intensity distribution particularly at the edges, and finally it disappears. To this must be added, that, at least in certain moths, the reduced glowing circle is surrounded by a black, and more peripherally by a brown halo (this is the phenomenon of the pseudopupil discussed later), and that the larger irregularities are probably due to the uneven illumination of different parts of the eye.

From the foregoing the reader will already have deduced the physiological basis underlying the extinction of the eyeshine. It is a result of the light-induced pigment migration discussed in Chapter V. For moths only the photomechanical action of the iris pigment is important. It is clear that when this has moved into the light-adapted position, not only is the retinal image (and hence the reflected light) much less bright, but also, with increasing light adaptation the glowing circle must reduce in size till it disappears. Thus (see Text fig. 19, p. 44), just as less and less light from the cones further away from the axis (*ab*) can reach the image point (*b*), so can fewer and fewer rays from the image point emerge through the lateral facets. Instead they will be intercepted by the proximally shifted iris pigment. Assuming the pigment migrates uniformly over the eye, then the rays furthest from *ab* will be the first to be blocked: the peripheral zone of the glowing circle disappears first, and with increasing light adaptation rays closer and closer to *ab* will be successively blocked: the circle shrinks until finally only a few, or even one facet allows light out: the glow vanishes.

This means that one can check the state of the iris pigment at any time in the living animal with the aid of the ophthalmoscope alone – a situation which has proved very useful for my studies on pigment migration.

The luminous pseudopupil also differs from the glowing vertebrate pupil in that the action of light on the iris pigment is local. The pupil in mammals simply contracts, maintaining its circular shape, whilst the luminous pseudopupil can assume different forms, light having influenced one part of the eye more than others. I do not dare to claim that the unilluminated parts of the eye remain entirely unaffected, but if they are affected at all then this could not be detected by the techniques I used.

One can show this local effect by examining an animal caught in the dark with the ophthalmoscope aligned in a particular direction. After the illuminated region has lost its glow, other regions still glow when the animal is turned so as to make them accessible to observation.

The following experiment gave the most striking demonstration of this effect. One evening a large convolvulus hawkmoth (*Sphinx convolvuli*) with glowing eyes was fixed to a strip of cork with the dorsal half of the eye looking at a lamp and some white paper, and the ventral half facing darkness. The two visual fields were separated by a sharp horizontal line. After ca. 10 minutes the dorsal half of the eye no longer glowed, whilst the ventral half still glowed beautifully. By suitable positioning of the animal one could focus on the boundary between the two halves: one then saw a perfect half moon, convex below, with a sharp horizontal boundary above. This thus corresponds to the situation represented in Text Fig. 19 where light can still emerge from the ventral half of the eye whilst in the dorsal half all the light is absorbed by the iris pigment. I have performed this experiment twice with the same result.

I also made the entire surroundings of the lamp and the room as dark as possible, so that the animal now only looked at the lamp with a group of facets. The result was that the whole eye still glowed, apart from a restricted area which had become dark. When the animal was rotated under the ophthalmoscope, this dark zone gave rise to peculiar optical effects as soon as the glowing pseudopupil approached it.

The disappearance of the pseudopupil is normally accompanied by a very noticeable change in colour. The initial red colour changes gradually to a dull orange and then whitish tone, in some species even ending as pale green. Unfortunately I do not have an adequate explanation for these changes.[54] It could, of course, be suggested that the original red colour originated from the visual purple in the retinulae and that this becomes bleached during illumination. However, at least in butterflies, I did not succeed in detecting pigment bleaching under the microscope, even in direct sunlight (see p. 105).

With respect to diurnal insects, my investigations have dealt mainly with butterflies. Many butterflies have a dark principal pseudopupil, visible with the naked eye, which is surrounded by a group of accessory pseudopupils. These all appear dark with or without the ophthalmoscope and will be discussed in detail later. The principal pseudopupil is usually easy to distinguish from the accessory pseudopupils by its darker tone and its central position. On turning the eye it moves like the luminous pseudopupil in moths. If one examines, for example, the cabbage white (*Pieris rapae*) with the ophthalmoscope using the sun as a light source, a very bright red spot can be seen in the centre of the principal pseudopupil. It is so small that not everyone would recognize it with the naked eye (I usually use a convex lens of $f = 2-8$ inches in combination with the ophthalmoscope), and initially one gains the impression that it lights up temporarily only to disappear again immediately. There are two reasons for this. Firstly one can only see the glowing point when the light is incident precisely along the axis of the observer's eye. The ophthalmoscope must thus be very accurately aligned, whereas in moths it suffices to have just some light reflected from the mirror into the eye. Secondly the red spot loses its colour after one, or at the most a few seconds of intense illumination, becoming whitish and inconspicuous so that it is easily overlooked.[55] Butterflies do not usually stay still during the investigation, which sometimes kills them (possibly because of the heat), and they make agitated movements of the head. Consequently the bright red spot lights up only to fade again at once and then reappears in a new position in the wake of more head movements.*

The explanation of this kind of eyeshine follows directly from the explanation of eyeshine in moths when one takes the anatomy into account. Thus, even though a tapetum of the sort we find in moths is not present, we know that such a tapetum is not necessary for eyeshine. In a section, e.g. of the cabbage white (Fig. 34), the positions of the pigment layers are very similar to those in a more or less light-adapted moth. Since such a moth still has a residual eyeshine, it is only to be expected that diurnal insects also show comparable glow phenomena. Indeed the bright spot in the principal pseudopupil of the cabbage white is really very similar to the last vestiges of glow in the moth's eye, where one also sees a dark pseudopupil surrounding a reduced and now punctate circle. It does not matter that the dioptric systems in moths and butterflies are different, since in each case the rays from one illuminated retinal element must return along basically the same route they took in reaching the retina. Admittedly, one cannot pinpoint the level at which the light is reflected back – be it at the basement membrane, more distally (perhaps from the tracheae surrounding the retinulae, more proximally – or even from all three levels – see Fig. 34). However, because the pigment sheath surrounds the retinula, all rays not directed towards the crystalline cone must obviously be absorbed on the reverse path. Thus only axial rays will enter the dioptric apparatus, and these will undergo the analogous refraction which the incident rays underwent.

Thus if we consider a cross-section of the retinula as an illuminated – and hence luminous – area, then its inverted image will be projected on or near the distal corneal surface (just as normally an inverted image of a small external object is projected onto the tip of the crystalline cone). However, because of the lens cylinder action the rays will diverge so little that their projection will cover only a very small region of the visual field. This is exactly analogous to the normal incident light path where only rays from a narrow segment of the visual field reach the retinal element of one ommatidium, these, however, having a relatively high intensity. The cross-section of the most distal part of a retinal element thus contains points which are conjugate with corresponding points in that segment of the visual field, and all rays passing through this section will reach these points. However, as we saw, all rays returning out of the eye must pass through this section if they are to leave the eye at all.

* I should mention that with the aid of the ophthalmoscope a most beautiful glow can be seen from the ocelli of those insects which have them.

Thus what we see with the ophthalmoscope is a complete affirmation of the processes described in the sections on dioptrics. It proves that light radiates from each facet only in a very strictly defined direction, and that this is conspicuously bright; basically the same conditions must also hold for the incident light. In a dark room, with a candle flame for illumination, the bright spot is seen to undergo small shifts when the mirror is slightly rotated. These shifts are in the direction of the incident light, for example a shift of the mirror image of the candle to the right results in the glowing of a group of facets which also lies further to the right.

The size of the glowing spot in the pseudopupil is a measure of the resolution of the retinal image – one can thus consider the spot exactly as the retinal image of the sun seen from the outside.[56] A retinal image can also be seen in the human eye with the aid of an ophthalmoscope.

As already mentioned, I have no explanation for the fading of the eyeshine in butterflies since I could not detect the disappearance of the red pigment in the retinulae. In an attempt I teased apart the eye of a cabbage white and found the red thread-like structures of the retinulae under the microscope.[57] I then illuminated the preparation with sunlight focused by the concave mirror, but even after a quarter of an hour no change could be detected in the intensity of the colour.

Following this general discussion of the phenomenon of eyeshine in insects I now wish to describe certain details and my findings from different species.

Firstly, I have demonstrated that the eyeshine is a purely dioptric phenomenon, as in vertebrates, and that there is no light production within the eye. It can already be safely inferred from existing descriptions that there are no luminous bodies like the luminescent organ in *Lampyris*, but there might be phosphorescent substances in the eye which produce light a short time after illumination. Even fluorescent substances do this, and these have been demonstrated in the human eye.

I constructed an apparatus (similar to Becquerel's phosphoroscope) which, with the help of a rotating cog-wheel, allowed a dark-adapted moth's eye to be alternately illuminated and observed. With rapid rotation this could be repeated many times within a second, so that any phosphorescence, even of only very short duration, would certainly have been noticed.

The humming-bird hawkmoth (*Makroglossa*) is a day-flying moth and, as we have seen, the fine structure of the eyes differs from that of both nocturnal moths and butterflies (Figs. 17, 63). Its eyeshine is also characteristic: with the ophthalmoscope one sees a conspicuously large, green shining spot within the pseudopupil. I observed this repeatedly during the day in diffuse or direct sunlight.

The tiger moth (*Euprepria*) is another diurnal moth. Here the glow is similar to that in butterflies, i.e. only a small bright spot.

The eyes of the red underwing (*Catocala nupta*) are large enough for the glow to be locally extinguished (by illumination with a lamp for ca. 15 minutes) whilst other parts still glow. I brought one specimen into a light-tight room during the day when its eyes were uniformly dark; after 2 hours the eyes had developed a most beautiful glow.

Initially I believed that only those butterflies which did not have black eyes showed luminous pseudopupils. Thus despite extensive effort I was unable to see any eyeshine in another species of *Pieris* (either *P. crataegi or brassicae*) with black eyes. The same is true for many other black-eyed insects, especially beetles. However, in a tiger moth, I later found a butterfly-like eyeshine despite its black eyes and since then have observed the phenomenon in several species of black-eyed butterflies, e.g. the chalk-hill blue (*Lycaena*).

According to my records the following butterfly species show a typical luminous pseudopupil: *Vanessa urticae, C-album* and *cardui*, the tigermoth (*Euprepria*), *Coenonympha pamphilus, Lycaena coryda, Argynnis latonia, Polias, Coleas hyale, Polyomatus phlaeas*, the cabbage white (*Pieris rapae*), and a *Hipparchia* sp. (*medea*?). All of these showed a red reflection, whilst *Pararge megarea, Melanargia galathea* and *Epinephele* showed a green glow. As already described, this is also the case for *Makroglossa*, and *Hesperia coma* behaves very similarly. In the day *Plusia gamma* shows only a pale spot within the black principal pseudopupil, but this transforms into a luminous pseudopupil in the evening. It is obvious to suggest that this green colour is the result of long exposure to sunlight but the situation is not so simple that this can be judged without further investigation.

I have, of course, looked for the presence of a luminous pseudopupil in other insects as well, and as expected found that it is a very widespread phenomenon. The only exception were the beetles. Even after keeping, e.g. *Hydrophilus piceus*, in the dark, (in which case the iris pigment adopts a position very similar to that in moths), no glow could be detected. This may be related to the massive and thick retinal pigment in the beetles, but whether this is the only reason remains to be seen.[58]

One sees a beautiful eyeshine, similar to that in butterflies, in the ventral half of the dragonfly eye. I have seen it in several species, but mention *Libellula vulgata* (Diplex) here as outstanding.[59] Under certain conditions a shining bright spot can still be recognized in the centre of the very small dark pseudopupil of the local great green bush crickets (*Locusta viridissima* or *caudata*).

Insects with little black pigment show a different sort of eyeshine. A housefly, for example, shows a rather large, poorly defined glowing spot with a much brighter spot standing out in the middle. The eyes of other dipterans are in general similar. In some, e.g. *Eristalis*, the central bright spot, which I consider the analogue of the luminous pseudopupil, is particularly well formed.

The difference between the two halves of the dragonfly eye are seen very clearly with the ophthalmoscope. The dorsal half shows the fly-like pattern whilst the ventral half is like that in the butterflies. To get an idea of the two kinds of glow imagine the large dark spot in Fig. 67 as completely glowing apart from a narrow marginal band, and the central black principal pseudopupil in Fig. 66 with a glowing spot.

If one turns a dragonfly under the ophthalmoscope one is left in no doubt that the pseudopupil moves much more rapidly in the dorsal eye half than in the ventral. If we recall that, in a certain sense, the pseudopupil can be considered as the retinal image seen from the outside, this means that the retinal image of an object with a particular angular velocity moves more quickly over the dorsal retina than it would in the ventral. Earlier I gave reasons suggesting that the dorsal part of the dragonfly eye is basically intended for motion detection and the ventral part for form recognition. The behaviour observed with the ophthalmoscope gives valuable support for this.[60]

The size and the indistinct nature of the retinal reflection in insects with eyes like those of flies is readily explained when one considers the greater light scattering in the retina due to the lack of black pigment. Earlier I discussed light scattering as a reason for the large luminous pseudopupil in moths. In the fly's eye scattering must be more pronounced, and in the absence of the black iris pigment, the margin of the glowing circle will not be so sharp. However, as is in fact seen, it is still to be expected that the actual retinal image should stand out particularly brightly as the centre of the indistinct spot.

2. Relative Position of the Corneal Reflection and the Pseudopupil

The luminous pseudopupil can be seen in many insects but not in all. The small size and weak intensity of the pseudopupil has been mentioned as one possible reason for its apparent absence, but in other animals another factor may play a role.

I have explained why the luminous pseudopupil is seen in the depth of those ommatidia whose axes more or less coincide with the eye of the observer. Thus, as a rule, it appears in that part of the eye where the observer's line of sight is perpendicular to the insect's cornea. However, the cornea itself forms a more or less spherical curvature which can function as a convex mirror. Just as the human cornea, it thus projects a mirror image from any light source. When using the ophthalmoscope this mirror image must again be seen when the observer's line of sight meets the insect's cornea perpendicularly.

Thus, a priori, one might predict that the pseudopupil should be invisible since it should be obscured by the strong corneal reflection. Indeed I believe this to be the reason why the luminous pseudopupil cannot be seen in many animals.

The question rather becomes, how is the pseudopupil visible at all? There are two reasons. Firstly in many animals the corneal reflection does not form an image. These are all those where the cornea is not a uniform curved surface (as it is e.g. in aquatic beetles and crustaceans in Figs. 16, 45–56), but is divided into facets, each of which forms a small convex surface (e.g. in lepidopterans Figs. 28–36). Such a cornea does not project a uniform image of the light source, rather each facet acts as an independent convex mirror; the microscopically small mirror images from these only form a shimmer over the eye which does not interfere with the observation of the luminous pseudopupil. If the radius of curvature of the single facets becomes very large, i.e. if the shape of the entire surface approaches that of a sphere, then a corneal reflection, similar to that from a smooth corneal surface again occurs (e.g. in dragonflies, Fig. 58). Hence one can already determine the nature of an animal's corneal surface with the naked eye.

Secondly, one can often see the luminous pseudopupil despite a strong corneal reflection for the following reason.

When discussing the dioptrics of the *Limulus* eye (p. 16) we saw that the anatomical axis of the cone is not always normal to the corneal surface (Text

Fig. 9). In this situation the optical axis of the ommatidium can also deviate substantially from the perpendicular to the cornea. In this arrangement we recognized a means of significantly extending the animal's field of view without resulting in a dangerously strong curvature of the cornea. This strategy is much more widespread than could be expected from cursory microscopic investigations, as proven by the more precise investigation of the pseudopupil with the ophthalmoscope. One then sees that the corneal reflection is very often shifted with respect to the pseudopupil and, as the anatomical investigation of *Limulus* and a few other species leads one to expect, this shift reaches a maximum at the edge of the eye. In certain large dragonflies and grasshoppers, for example, the coincidence of corneal reflection and pseudopupil would have rendered the latter invisible if the two were not shifted relative to each other, away from the centre of the eye. The two optical phenomena are then spatially quite separate with the corneal reflection lying peripheral to the pseudopupil; indeed, the former can already have disappeared at the edge of the eye whilst the latter is still visible.

The explanation for this is obvious. One only needs to recall the reversibility of the ray path in a dioptric system. Then in Text Fig. 9 (p. 17) the ray from the illuminated point a' will follow the path $a'oc$, since, as we saw, the ray from c followed the route coa'. Therefore if the retina is illuminated with the ophthalmoscope at a', the luminous pseudopupil will be seen in the direction co; however, the corneal reflection will be seen at x, or at an even more peripheral point, where the corneal surface is perpendicular to the observer's line of sight.

In the centre of the eye (e.g. cone A in Text Fig. 9) the pseudopupil and corneal reflection coincide, since the path of the ray reaching a is perpendicular to the corneal surface.

In a theoretical respect the spatial mismatch of pseudopupil and corneal reflection thus gives a striking confirmation of the broadening of the visual field which had been discovered by independent means. In a practical sense it is important because it still allows the pseudopupil to be seen in animals where a strong corneal reflection would otherwise have made it impossible.

3. *Crustaceans*

The occurrence of eyeshine amongst the crustaceans is in fact only to be expected when one bears in mind the similarity in the structure of insect and crustacean eyes. However, as far as I know eyeshine has not been previously described in crustaceans, presumably because of the difficulty in obtaining fresh material.

With the aid of the ophthalmoscope and sunlight I first saw a red luminous pseudopupil in a dark-adapted specimen of *Astacus fluviatilis*, the crayfish commonly found in our lakes and streams. As the glow faded a black pseudopupil remained. I have performed the experiment using both sun and artificial illumination, in the evening and the morning, generally speaking with the same result. Occasionally the eye appeared dark when I had expected it to glow (but never the other way round). I now know that this is related to the animal's condition. In crustaceans the development of eyeshine is much more sensitive to the animal's health – as far as this can be judged – than in insects. I have already pointed out this fact with respect to the migration of the iris pigment, which is, of course, causally related to eyeshine.

Despite having a retinal tapetum, the eyeshine in *Astacus* is much less pronounced than in some marine crustaceans. Figures 71 and 72 show the eye of a hermit crab (*Pagurus*); Figs. 69, 70 that of *Palaemon*, whose glow is as bright as that in moths, and Figs. 73, 74 the eyes of *Sicyonia sculpta*. In each pair of figures the dark- and light-adapted eyes of the same animal are shown next to each other. The glow in the dark-adapted eye disappears in a matter of minutes.

I saw a similarly intense eyeshine following dark adaptation in *Dromia vulgaris* (note also the extreme dark-adapted position shown by the iris and retinal pigment in this animal – Figs. 55, 56), and also in *Nica edulis* and others. Whilst in these animals the eyes show a large glowing disc, as in moths, in other crustaceans the situation approaches that in butterflies in as much as the luminous pseudopupil is smaller and seen within a black pseudopupil, or else the glowing area is less intense and not so well defined. Nevertheless, there is always an unquestionable glow that rapidly disappears in a light room. Without giving more details I have seen the latter type of eyeshine in *Carcinus maenas* (Fig. 68), *Innachus, Scyllarus arctus, Pisa, Galathea strigosa, Herbstia* and *Maja verrucosa*. On illumination the glow disappeared completely in *Innachus, Galathea, Pagurus*, the various species of *Palaemon*, and *Sicyonia sculpta*, whilst in *Carcinus maenas* a small residual luminous pseudopupil remained, as in butterflies. It is known that *Carcinus* behaves as a diurnal animal, also being active in sunlight.

Palinurus (spiny lobster), *Homarus* (true lobster) and *Scyllarus arctus* (the bull-dozer shrimp) are peculiar in that in the dark their eyes show an obvious extensive red glow, without sharp boundaries. In the light the glow becomes weaker, also changing colour, but still remains as an extended whitish shimmer. Because of the poorly defined boundaries one cannot speak of a pseudopupil.

With the ophthalmoscope, I could see no glow at all and no light-induced change in *Portunus*. Subsequent histological investigation showed, in fact, that this crab has no tapetum, although the iris and retinal pigment migrate on illumination (Figs. 37, 38). The scarcity of reflected light presumably accounts for my inability to see an optical effect of this migration.

In *Peneus membranaceus* and *Squilla mantis* as well, the eyes of specimens taken out of dark containers had the same appearance as those kept in the sun. In these species the optical investigations were perfectly in accord with the anatomical studies, as no differences were found between sections taken from light- and dark-adapted eyes. As we have seen, *Peneus* has practically no pigment.

As in insects, in almost all the crustaceans I have investigated, the corneal reflection and pseudopupil do not usually coincide at the edge of the eye, the corneal reflection again lying more peripherally. Indeed, as a rule, one can only see the pseudopupil well when one seeks out such eye regions – at least when the pseudopupil is small. Thus, here as well, the ommatidial axis deviates considerably from the perpendicular to the corneal surface, and, as has been discussed for *Limulus* and certain insects, the visual field in these crustaceans is greater than suggested by the corneal curvature.

The following observations on the physiology of pigment migration are presented more in the hope of stimulating further investigation than to provide a finished account.

The transition from the dark-adapted to the light-adapted state is much more rapid than vice versa. Initially, I had believed that crustacean eyes glowed only at night, as I usually only saw them glowing in the morning when I took them out of their darkened containers. Later I found that the glow is also present during the day provided the animals are kept long enough in the dark. The glow disappears within minutes in sunlight, and reappears only after hours in the dark. I also found, e.g. in various species of *Palaemon*, that even if the dark container was opened only briefly (so that the eye still glowed when it was closed) after another hour the eyes no longer glowed. Only after 3 hours did I again find a conspicuous eyeshine.

An obvious question is whether the pigment migration is a direct result of illumination, or whether the movements are a reflex action. I have tried to answer this question with a variety of methods, but have not achieved a satisfactory result. In a number of animals (a large *Palinurus*, two *Scyllarus*, three *Maja* and one *Pagurus*) I made a hole in one side of the eyestalk and removed all the soft parts including the optic nerve. The two eyes then did indeed show differences in their reaction to darkness or illumination (as observed with the ophthalmoscope) but the results were not convincing. The only conclusion from the experiment was that the operated eye reacted either not at all, or much less than the intact eye. This can obviously also be attributed to the disrupted supply of nourishment.

I have also made a whole series of experiments (*Palaemon* was best suited for this) in which I cut off one eye (including its stalk) from animals kept in the dark, and then brought the animal and eye into the sunlight (and vice versa). In the former case the eye remaining on the animal naturally stopped glowing, whilst the amputated eye continued to glow. In one case I found the eyes (I performed the experiments on several animals at once) still glowing in the sun the next morning, 17 hours later. Conversely, I removed one eye each from several *Palaemon* which had been kept in the evening sun and brought both eyes and animals into a dark room; the next morning I found the intact eyes to be glowing whilst the amputated eyes were still in the extreme light-adapted position. Naturally the amputated eyes were protected from drying out.

These experiments indeed support the hypothesis of reflex movements, but are by no means without objection and were only intended as pilot studies to check the experimental possibilities. Unfortunately, I did not have enough time in Naples to answer the question definitively. It is obvious that a sufficiently rapid loss of excitability following removal of the eye could also result in the eye remaining in the state it was in at the moment of extirpation.

I have repeatedly emphasized the importance of the animal's health for normal pigment migration. It happened, for example, that one crustacean out of a whole batch kept in dark containers jumped out of the aquarium. Although it landed in a cup containing water a few centimetres deep it was the only specimen with no eyeshine, although otherwise one could detect no ill-effects.

Finally, I attempted to see if pigment migration could be induced by electrical stimulation. In this way one should find out which is the resting state of the pigment. Unfortunately these experiments, performed mainly on *Hydrophilus* in which one can only judge the position of the pigment by histological investigation, also yielded no satisfactory results. One could have expected the best results from the stimulation of glowing and non-glowing eyes of a large moth. However, I obtained these only during my stay in the country where the equipment for electrical stimulation was not at my disposal.[61]

B. The Phenomenon of the Pseudopupil

I come to the discussion of a remarkable optical phenomenon, the unravelling of which has cost me much puzzlement. As far as I know, it was again Leydig (1855, p. 431) who gave the first accurate description, otherwise the subject has hardly been discussed apart from a short remark by Thompson Lowne (see below). The general occurrence of the phenomenon – it can be seen in the majority of compound eyes – and the common basis underlying the variety of its numerous forms have certainly not been previously pointed out, and an explanation hardly attempted.

Leydig correctly says that one sees a dark spot resembling a pupil in the *Limulus* eye, which differs fundamentally from the vertebrate pupil in that it changes its position with that of the observer. He already gives it the name "pseudopupil". He also points out that one can sometimes see several such black spots in the eyes of butterflies. Thompson Lowne (1884) noticed six black spots surrounding a centrally located seventh in the cabbage white. These changed position with the observer and appeared bright when seen with an ophthalmoscopically equipped microscope. (I have no doubt that he was in error here and that only the central spot appeared bright, i.e. the phenomenon described in the previous section).

In the previous section I discussed the luminous pseudopupil. In the following I ask the reader to ignore this phenomenon, which corresponds to eyeshine in the vertebrates, and rather to imagine a pupil-like dark spot.

As a rule pseudopupils are seen in compound eyes with a light-reflecting substance between the distal parts of the crystalline cones (iris tapetum). Usually this consists of pigment directly behind the cornea. This pigment can vary greatly in colour and brightness; the brighter the colour the more pronounced is the pseudopupil. Thus one sees pseudopupils against red (Fig. 65), blue (Figs. 71, 76), yellow (Fig. 61), blue-green (Figs. 66, 71, 75), light brown (Figs. 64, 68), brown (Figs. 62, 63, 73) and dark brown (Fig. 73) backgrounds, which can finally be so dark that one can hardly see the pseudopupil at all, e.g. in many non-glowing moth eyes, *Hydrophilus piceus* and *Dytiscus*.* (Similarly the human pupil can be hard to distinguish when the iris is very dark). In insects in which the most distal pigment layer is black (mainly beetles) one no longer sees any pseudopupil. I do not know of any crustaceans of this sort.**

* With these dark eyes one must use strong illumination with the ophthalmoscope in order to convince oneself of the presence of pseudopupils.

** In the spiny lobster and the true lobster the pseudopupils are indistinct for other reasons.

All pseudopupils have the property of moving with the observer, and in general they are to be seen where the compound eye is intersected perpendicularly by the observer's line of sight. The frequent exception to this rule, particularly at the eye margin, has already been discussed in detail and explained for the luminous pseudopupil.

The pseudopupil is by no means always round. Corresponding to the deviation of the corneal surface from a spherical form, it also often takes on an oval form which, in insects with hexagonal facets, sometimes tends to appear as an irregular hexagon. In crustaceans with square facets, at least with moderate magnification, it usually appears as an obvious rectangle (Figs. 65, 69, 73). The crayfish also shows a pseudopupil. The beautifully coloured eye of *Peneus membranaceus* has, as we saw, virtually no pigment, nevertheless, thanks to its iris tapetum, it shows a clear square pseudopupil arising from the crossing of two meridional bands (Fig. 75). Correspondingly it also has square facets.

The characteristic corneal curvature in some crustaceans, e.g. *Carcinus maenas*, results in a tapered appearance of the pseudopupil (Fig. 68).

Apart from these pseudopupils, in many species one also sees other dark spots in the eye. These are not so dark and are less well defined, but nevertheless have much in common, in particular moving their position with that of the observer. After close examination of several animals, particularly butterflies, one soon recognizes that there is in fact an order to the apparently irregular group of dark spots. Surrounding the pseudopupil proper, first comes a crown of six dark spots which approach the pseudopupil in clarity, then a further crown of twelve less distinct spots. In certain dragonfly larvae I have also distinguished spots belonging to a third crown.[62] These all belong to the phenomenon of the pseudopupil which thus consists of the principal pseudopupil – till now the sole topic of discussion – further (in species with hexagonal facets) six first order accessory pseudopupils, probably twelve second order accessory pseudopupils, and finally third order ones, all of which move across the eye when it is turned.

Rather like the phenomenon of the rainbow, which is seldom seen in its full splendour, I believe that most eyes do not reveal all the pseudopupil phenomena for secondary reasons.

The phenomenon is seen most completely in late instar aquatic larvae of *Agrion*, the eyes of which are studded with spots. It was here that I first noted the arrangement described and the numerical distribution. If one turns the eye so that the principal pseudopupil moves towards the margin, the accessory pseudopupils on one side disappear as they reach the margin, whilst on the opposite side new accessory pseudopupils find space to appear. This is true for all animals with any accessory pseudopupils at all. The ventral eye half of large dragonflies (Fig. 66) also demonstrates the phenomenon rather well when one brings the principal pseudopupil to the correct position. One then sees the distorted, partially fused first order accessory pseudopupils, on one side an almost perfect row of second order ones, and occasionally at isolated points on the margin one or other of the third order ones can be seen (Fig. 66 at *a*). Any movement of the animal or the observer results in a flowing pattern which gives the eye its mysterious shimmer. This becomes even more confusing when the principal pseudopupil approaches the dorsal eye region where it eventually becomes half-moon-shaped and transforms into the large black principal pseudopupil of the dorsal eye with the loss of all accessory pseudopupils.

The phenomenon can still be well seen, though less extensively, in the eyes of many butterflies, e.g. in the cabbage white (*Pieris rapae*, Fig. 61). In a given position of the eye one can see a fine principal pseudopupil, six well-developed first order accessory pseudopupils and some indications of the second order ones. The latter are more conspicuous on one side when one shifts the principal pseudopupil to the other. The situation is very similar in the brimstone (*Colias rhamni*).

Epinephele also shows second order accessory pseudopupils. These tend to merge with one another and with the principal pseudopupil, giving the eye its characteristic appearance (Fig. 64); only after inspection from various angles can the regularity of the arrangement be recognized. When looking at these figures one should not forget that each represents an eye seen from a particular direction. With only a small shift the position, shape and relative spacing of the pseudopupils changes.

In the painted lady (*Vanessa cardui*, Fig. 62) the pseudopupils are similarly spaced though differently arranged. Here there are indications of second order accessory pseudopupils. In *Melanargia* and *Makroglossa stellatarum* the phenomenon has taken on greater dimensions, so that even the first order accessory pseudopupils do not quite fit within the eye (Fig. 63). In the latter the principal pseudopupil is relatively large. It was this pseudopupil which glowed such an intense green in the ophthalmoscope (p. 105). If – as claimed by Thompson Lowne – the accessory pseudopupils could also glow I would certainly not have missed it in this moth.

If one imagines the principal pseudopupil as even larger and the accessory pseudopupils as having shifted out of the eye altogether, then we have the situation which, as mentioned above, is found in the non-glowing eyes of moths and certain beetles.

The large grasshoppers, e.g. *Psophus stridulus*, have sharp, but small principal pseudopupils but no accessory pseudopupils.

In all these cases, as shown in the figures, the ring-shaped space between the principal and first order accessory pseudopupils is particularly bright. In some animals, e.g. *Hesperia coma*, which has an eye very similar to that of *Makroglossa*, this ring of light is so pronounced that one is inclined to say the principal pseudopupil is surrounded by a six-armed, whitish star.

A conspicuous difference between the principal and first order accessory pseudopupils on the one hand and the second and third order accessory pseudopupils on the other, is that the position of the former is dependent only upon the direction of observation, whilst the position of the latter is also dependent upon the direction of illumination.[63]

I have observed this in *Pieris rapae* with the aid of a magnifying glass. Using a candle flame as a light source in an otherwise darkened room, one first identifies the principal pseudopupil with the ophthalmoscope (by finding the glowing red centre). Then if one turns the eye so that at least some second and third order accessory pseudopupils are visible, one sees how these move in an arc around the principal pseudopupil when one moves the candle up and down. I have never been able to induce a noticeable movement of the principal or first order accessory pseudopupils in this manner.

The principal pseudopupil is thus fixed in that its position is independent of the direction of illumination. However, the same is not true for its central glowing spot. At least in *Pieris* – and this will certainly be the case in other butterflies as well – this "luminous pseudopupil" undergoes small movements within the black principal pseudopupil when the direction of illumination from the ophthalmoscope is altered. I have already pointed this out above (p. 104) and provided an explanation.

As in insects, the dark principal pseudopupil in many crustaceans is also surrounded by a conspicuously bright ring (e.g. *Carcinus maenas* Fig. 68).

Despite the clarity of the, albeit small, principal pseudopupil, I have never seen unmistakable accessory pseudopupils in crustaceans except in the hermit crab (*Pagurus*, Fig. 71). Like insects, this crab has hexagonal facets, and probably also has six accessory pseudopupils. They are very indistinct, can only be seen under the most favourable conditions, and I never succeeded in counting them.

In crustaceans with square facets, where one would expect to see four first order accessory pseudopupils, I have never seen any with such clarity as in insects. The four darkly coloured zones in, e.g. *Galathea* (Fig. 65), or the peculiar pattern in *Palaemon* (Fig. 69) and similar phenomena in *Palinurus* and *Homarus* I consider to be caused by reflections from the rows of corneal facets (or the underlying cones); they thus have a different origin from that of pseudopupils. The arms of the dark cross in *Peneus membranaceus* (Fig. 75) might still be identified as pseudopupil phenomena.

1. Explanation for the Pseudopupil Phenomenon

Even though I have to restrict myself to the optical mechanisms responsible for the principal and the first order accessory pseudopupils, the explanation for the phenomenon is not simple. The higher order accessory pseudopupils are too poorly developed for a more detailed physical investigation, although a similar explanation probably holds.

The great variety in the form and size of the phenomenon in the numerous different compound eye types forces us to seek an explanation for a typical case in a typical eye and then to consider the individual variations on the basis of the underlying anatomical situation. Thus initially we assume a perfectly regular eye (i.e. spherically curved with perpendicularly arranged cones).

The phenomenon is disposed around a centre, the middle of the principal pseudopupil, where the corneal surface is perpendicular to the observer's line of sight. This line will be called the axis of the phenomenon. As far as we intend to explain it, the phenomenon consists of the principal pseudopupil centred on the axis, and six accessory pseudopupils equidistant from the axis and arranged in a hexagon around it.

2. *The Principal Pseudopupil*

As noted by Leydig this is seen very well in *Limulus*. I could even recognize it in a dried specimen. Although there are no accessory pseudopupils in this animal it is particularly suited for the study of the principal pseudopupil not only on account of its size – it is several millimetres in diameter – but above all because of the size of the individual facets. If one looks at the edge of the pseudopupil in the living animal with a magnifying glass one can see the light distribution in the individual facets (Fig. 8). *A* lies within the pseudopupil region, *B* outside it. The pseudopupil is thus black because near its axis no light from the facets reaches the observer's eye, and the boundary of the pseudopupil is determined by facets further from the axis from which light (in the direction of the observer) does emerge. The figure also shows that each of these facets is bright on the side away from the axis, and that the increasing brightness of the eye away from the axis results from the bright half-moon-shaped zone in each facet increasing in width and intensity.

In animals with accessory pseudopupils (e.g. dragonfly and cabbage white), the light distribution in and around the principal pseudopupil is the same as in *Limulus*. As already clear from the foregoing, the direction of illumination is generally speaking unimportant.

The reason for this light distribution and thus the origin of the principal pseudopupil follows from Text Fig. 23. It is important to grasp that light from a point source is only incident perpendicularly at one position on the eye's surface, and for other corneal facets the angle of incidence increases with the distance from this position. For the sake of convenience let us imagine the observer's eye as a point source and ask what is the fate of the rays in the compound eye? If we find the path of these rays then, from the reversibility of light, we also know from which points in the compound eye light can reach the observer's eye.

A and *B* are two dioptric elements,the shape of which is taken from the *Lampyris* eye since I am most familiar with this eye. In other eyes the behaviour must be basically the same. If the eye of the observer lies on the line *ba*, i.e. on the axis of the phenomenon, then light from it would enter the facet *B* and illuminate an axial position in the retina. If this position is not cloaked in black pigment but can reflect sufficient light, then this will reach the observer's eye along *ba*. The centre of the principal pseudopupil – as described above for butterflies, dragonflies and certain crustaceans – will thus appear bright. If no reflecting layer is present then no light will return to the observer's eye and the centre will appear black as discussed, e.g., for beetles and light-adapted crustaceans.

If, however, a ray from the observer's eye strikes a facet obliquely, e.g. *co* – the facet would then in fact be some distance from the axis of the phenomenon – then, as we saw (p. 16) it reaches the point *m* and if, as in almost all eyes, black pigment is present here it will be absorbed. Therefore, from all facets with a certain inclination to the axis only those rays originating from black pigment can reach the observer's eye. In other words, no rays will reach his eye; instead he sees the black pigment surrounding the cones in the form of a ring centred on the luminous pseudopupil (Fig. 72), or (when the retina is non-reflective), as the outer zone of the dark spot resulting from the rays discussed first. Both together then

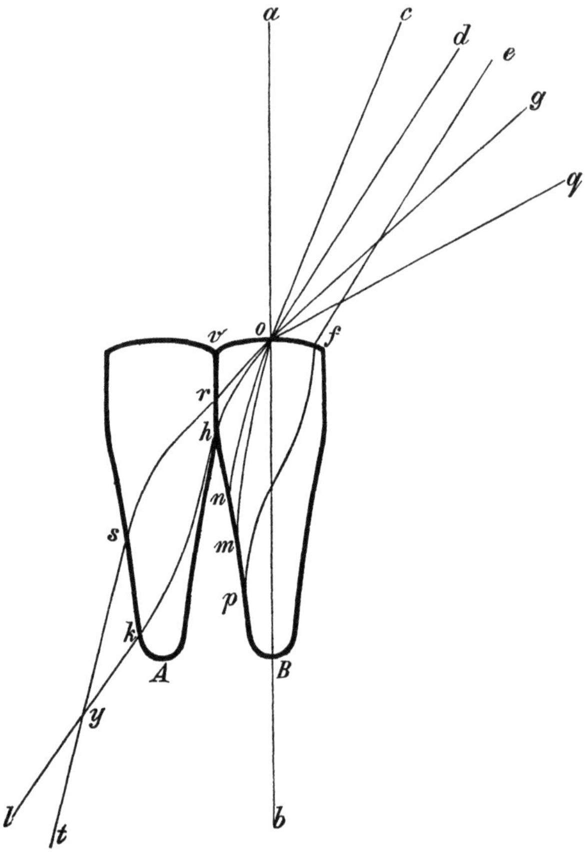

Text Fig. 23

form what we have called the principal pseudopupil. This outer black ring is always present whatever the conditions of illumination since the pigment which causes it absorbs all incident light.

A ray from the observer's eye striking a facet with an even greater inclination, e.g. *do*, reaches the pigment lying at the point *n*. We have seen that in many compound eyes there is a less dense, often beautifully coloured pigment lying distally to the black pigment – the iris tapetum – e.g. the yellow pigment in the cabbage white (*T*, Fig. 34). Thus ample light, in this example yellow, reaches the observer's eye, since from all the facets with the appropriate inclination to the axis of the phenomenon, it is only this yellow light which can proceed in the direction of the observer's eye. This thus determines the outer boundary of the principal pseudopupil and also gives rise to the bright halo which one usually sees around it (Figs. 62, 64).

The light distribution in single facets described above (see Fig. 8) is also simply explained in this way. Thus it is clear that a ray *ef*, which strikes the facet at *f* from the same direction as *do* (i.e. also coming from the observer's eye), will hit the cone margin at *p*, i.e. in the region of black pigment. Thus the corneal facet will appear black on the side facing the axis of the phenomenon whilst the other side is already bright.

This interpretation implies that a pseudopupil can be seen only in eyes where the presence of a transparent, or at least not completely opaque, substance distal to the black iris pigment . In this respect my experience agrees entirely with the theory. This is thus the reason why one sees no pseudopupil (apart from the luminous pseudopupil seen with the ophthalmoscope) in lepidopterans with black eyes and also in most beetles. Of course, many do not have completely black eyes, but rather dark brown; the corresponding pigment then has the same relative position as the iris tapetum in other animals (*c*, Fig. 15). However, then one also sees the principal pseudopupil. *Limulus* forms an apparent exception since one sees a conspicuous pseudopupil although the only pigment in its eye is black. Here, the function of the iris tapetum is effectively replaced by strongly reflecting layers of corneal material found between the cones. Thus numerous, extremely fine, pore canals penetrate right through the cornea and appear to be filled with air (*a*, Fig. 20). It is clear that this modified chitinous substance, arranged around the distal parts of each cone (in the functional sense) not only can replace the bright pigment of other eyes in terms of its position and reflecting ability, but can even surpass it.

That the dorsal part of the dragonfly eye still shows a large diffuse pseudopupil even where there is no black pigment is because the brown pigment in question is still very dense. The diffuse and extended glow of the pseudopupil is in fact due to the pigment not being entirely black. In the ventral eye the pseudopupil is black and sharply defined because here the pigment is black and the the cones are relatively long and narrow compared to those in the dorsal eye (see above, Chap. VI and Fig. 59).

In the foregoing I considered the origin of the principal pseudopupil only in terms of the pigment surrounding the cones. However, in those animals with retinulae reaching up to the cone tips, there is no doubt that the surrounding retinal pigment must also function in the same way. With respect to the pseudopupil it then simply forms a proximal extension of the iris pigment.

Superposition eyes with very extensive pigment migration require special comment. As best seen in the eyes of moths in the daytime, there is a conspicuously large pseudopupil weakly resolved against a brownish surround. Where is the necessary distal pigment layer?

Such a differentiated pigment layer is in fact present though it differs only slightly in colour from the iris pigment in histological preparations. Nevertheless, it is unquestionably dark brown. It also has a different physiological behaviour from the iris pigment. Thus it forms a thin layer closely adhering to the cone which, even in the dark-adapted position, is easy to distinguish from the iris pigment proper. At least in some cases (e.g. in the red underwing, Fig. 15) it maintains its position on the cone during light adaptation. In crustaceans with extensive migration of the iris pigment, this pigment and the iris tapetum (the analogue of the less dense pigment layer in insects) move together without changing their relative position (see Figs. 39, 51–54). In other crustaceans with less extensive migration, the iris tapetum does not move (e.g. in *Portunus*, Figs. 37, 38 *i.t.*). In many moths, as in most insects and crustaceans with superposition eyes, the cone margin is more or less devoid of pigment in the light-adapted state. One might now ask how does a pseudopupil arise when the rays *co, do* and *ef* (Text Fig. 23) do not encounter any pigment at all? The answer emerges when we consider the normally refracted rays in the eye. In these animals, where the dioptrics are equivalent to a lens cylinder twice as long as its focal length, rays with a moderate inclination to the axis will still be guided to the cone tip where they will emerge on the same side of the axis as they entered. This has been discussed in detail above (see Text Fig. 12 p. 30 ray *pb'p'*). However, these rays also reach the proximally migrated iris pigment where they are absorbed. Thus the entire area of these obliquely incident rays belongs to the principal pseudopupil although their path is quite different from that described up to now. The observer then sees a black pseudopupil which does not arise from the sides of the cone sheathes away from the axis of the phenomenon, but from the sides of the pigment sheaths lying between cone and retina and facing towards the axis. Once they have left the cone, rays which proceed as *co* or *ef* (Text Fig. 23) must sooner or later encounter black pigment, which has now shifted far proximally.

Thus, all else being equal, the larger the number of facets involved in the image formation of one point in the dark-adapted eye, the larger will be the pseudopupil in the light-adapted state. This also seems to agree with experience.

3. *First Order Accessory Pseudopupils*

An understanding of this part of the phenomenon is gained principally from two observations.

1. The eye of a living animal with well developed accessory pseudopupils (e.g. cabbage white, ventral eye of dragonfly) is examined under the microscope with incident light at a magnification allowing the hexagonal facets to be well seen. One then notices that each accessory pseudopupil lies on a line from the centre of the principal pseudopupil, which intersects opposite sides of the hexagonal facets perpendicularly (i.e. corresponding to the short diameters of the facets). If we consider a plane divided into regular hexagons, it is clear that through the centre of any one, one can draw three lines that intersect the sides of all the hexagons which they meet, perpendicularly. These form the three axes along which the accessory pseudopupils are to be found.

However, the eye is curved and since in general a perfectly regular distribution of hexagons of a given size is not possible on a spherical surface the directions of these axes deviate, often considerably, from those on the hypothetical surface. This is the reason why the six first order accessory pseudopupils are usually far removed from a regular array, i.e. by no means at equal distances from each other or from the centre of the principal pseudopupil. This is amply demonstrated in the figures on Plate VI.

Although it seemed apparent from the outset that the number of first order accessory pseudopupils was causally related to the number of sides to each facet, this supposition is reinforced with this relationship of their positions.

2. The second observation concerns a very surprising optical phenomenon; namely the appearance of what I shall call accessory images.

After brushing out the eye of *Lampyris spl.*, under suitable conditions one sees six additional upright images around the normal (principal) retinal image. They are basically similar to the principal image but less sharp. As with the principal image one can also see the rays converging on the corresponding image point when using a point as the object. It is then very obvious that the rays converging on each image point derive from only one sector (the six sectors being divided by the three intersecting axes mentioned above) so that rays emerging from cones lying near the demarcation lines diverge from one another. The accessory images are arranged just like the first order accessory pseudopupils in other eyes, and are similar in that: usually one cannot see all six simultaneously, the distance between them and from the principal image are subject to small variations, and they again lie on the three axes, passing through the centre of the principal image and intersecting the sides of the hexagons perpendicularly.*

The accessory images are still clear enough to distinguish the fingers of a hand held nearby, especially when it is moved. They lie in a somewhat different plane from the principal image but are approximately the same size. I have seen them not only in *Lampyris* and those relatives with fused cone and cornea, but also in *Tropinota hirtella* and its relatives. I emphasized previously (p. 54) that the major dioptric function in these animals is performed by the corneal facets, and thus, even when the inconspicuous cones have been brushed away together with the pigment, one still sees an, albeit poor, upright retinal image, along with the accessory images, all arising in a very similar way to those in *Lampyris*.

These accessory images now allow one to trace the ray path within the eye, and thereby make inferences about the accessory pseudopupils.

Starting with the accessory image of a point source in *Lampyris* and then focusing down towards the object, just as with the principal image, one notices that the ray bundles (which are of course less sharply defined and strongly distorted) diverge from each other and each merges into a cone some distance from the axis. One then sees a circular group of bright cones, similar in number to those contributing to the principal image. One can now see that all the light contributing to the accessory image emerges from the marginal surface of these cones. In vivo, however, this marginal surface is cloaked with iris pigment,** which thus absorbs the entire accessory image so that it is meaningless for the animal's vision.

If we now return to our concept of the observer with a luminous eye, then it is clear that the rays leaving it must be absorbed by the iris pigment in these six groups of cones, specifically on the side away from the axis of the phenomenon. We have no reason to assume a similar absorption in the space between these six

* In the *Lampyris* eye one cannot see the hexagonal facets in the arrangement described above; in order to see the position of the accessory images relative to the sides of the hexagons, I used the trick of turning the preparation over as it was. One then also sees, as described on p. 26, an image in the plane of the retina and also the accessory images. However, now one can also focus on the hexagonal facets and thus learn the relative position.

** Or in the light-adapted position the pigment lies deeper and will also catch the emerging light, only somewhat later. In *Tropinota* the light emerges through the lateral walls of the inner corneal surface.

groups so that apart from the principal pseudopupil the observer must also see six other black spots on the eye: the six first order accessory pseudopupils.

Therefore the existence of accessory pseudopupils follows necessarily from the existence of accessory images, and both share the same physical explanation. What, then, is the origin of the accessory images?

Let us consider parallel rays incident on a spherically curved eye. The ray incident normal to the eye's surface is then the axis of the pseudopupil phenomenon. Next to the axial facet the incident rays form an angle with the ommatidial axes which increases with the distance from the axial facet. The rays entering the eye (*com* in Text Fig. 23) will first be absorbed by iris pigment, then (*don*) reflected (also partly absorbed) by the light-coloured pigment, thus giving rise, as we saw, to the phenomenon of the principal pseudopupil and the surrounding bright halo. Even further from the axial facet the angle of incidence is even greater (*go*) and the ray no longer impinges on the light-coloured pigment, but passes between the dividing plane of two corneal facets at *h*. It thus reaches the neighbouring cone where it follows a path which will be given approximately by *ohkl*. These are the rays we see emerging from the cones at *k* in the accessory image. Since iris pigment is present at *k*, it is clear that the facets of this part of the eye must appear black to the observer. Since there must be a ring of ommatidia with the necessary inclination to the axis, one might expect to see a black ring. In fact, in certain animals one sees unquestionable indications of this, e.g. in *Epinephele* (Fig. 64). However, generally speaking, not only does this ring tend to separate into six parts, as in this animal, but it really does consist of only six spots, the accessory pseudopupils or images. This separation is apparently due to the corneal facets.

A contributory factor can readily be found. We saw that the accessory pseudopupils lie on lines intersecting the hexagonal facets on their short diameters. Thus if *vf* (Text Fig. 23) represents a section through the short diameter of the corneal facet, then the distance *vf* would be greater if the section was through the long diameter. The refracted ray *oh* would not then reach the neighbouring cone but would (like *on*) reach the light-coloured pigment between the cones. This explains why the halo around the principal pseudopupil is narrowest along the three axes intersecting the facet sides perpendicularly, thus forming a six-armed star which defines the inner boundary of the first order accessory pseudopupils.

The peripheral boundary of the phenomenon is equally simple to explain. As stated, the accessory images show that rays such as *go* leave the cone margin at *k*. A more strongly inclined ray, *qo*, will be refracted approximately as shown by *orst*. At *s*, however, there is again the light-coloured pigment, which thus gives rise to the outer boundary of the first order accessory pseudopupil.

So far one might have expected the accessory pseudopupils or images to form an undulating ring surrounding the principal pseudopupil. Each of the six outwardly directed peaks would correspond to an axis along the long diameters of the facets. That the phenomenon is not a continuous ring, but divided into six spots, is because the light rays passing the boundary between two facets (at *r* or *h*, Text Fig. 23) undergo a sideways deflection unless the boundary lies on one of the six preferred meridians (corresponding to the three axes mentioned previously).

This, and the origin of the accessory images, is treated in the following. Let us consider a meridian plane defined by the axis of the phenomenon and one of the meridians intersecting the sides of the facets perpendicularly. Rays parallel to this plane, as long as they are incident on facets where they will be refracted between the facet boundaries, will converge approximately to a point lying in the meridian plane. In the whole eye there are six such convergence points, the accessory images. Rays actually in the meridian plane converge since they pass through dioptric components which are basically lens cylinders. These will be obliquely penetrated, and from what we know about lens cylinders, the ray paths in Text Fig. 23, *gohkl* and *qorst* should be pretty accurate. Indeed the direction of the emerging rays has been experimentally observed. The diagram shows that the rays entering a facet are deflected in an arc. With respect to the ray path following emergence, the shape of the cone and the point of entrance into the cone naturally play a role. Direct observation shows that the emergent rays cross over.

Rays such as *st* and *kl*, but coming from different ommatidia in the meridian plane, thus converge behind the dioptric apparatus, giving rise to the possibility of accessory image formation. However, it is also necessary for rays in facets outside the meridian plane to be deflected towards the meridian plane.

If one draws hexagons as regularly as possible upon a spherical surface, one sees – as in the insect eye – that at some distance from the line formed by the intersection of the meridian plane with the eye's surface, it is no longer the sides of the hexagons but their corners that are directed towards the axial facet. This is also naturally the case on a plane. However, each side of the hexagon represents the boundary surface between two corneal facets in which the points *h* and *r* (Text Fig. 23) lie. A perpendicular to one of these boundary surfaces falls in the meridian plane when the facet is on the meridian, otherwise it is inclined to the meridian plane. If we consider the facet boundary to be struck by a perpendicular ray, then this ray must remain in the plane given by the ray and the perpendicular to the facet boundary, whatever refraction it undergoes at the boundary surface. Since this perpendicular plane intersects the meridian plane, the ray will be guided towards the meridian as its refraction in the cone is otherwise similar to that of rays which are incident on the meridian plane itself. From this it would be quite conceivable that two boundary surfaces of a single corneal facet would separate the light passing through them so that one would direct light to one accessory image and the other to another one. However, this is not true for the animals which I have investigated; between two groups of cones which contribute to accessory image formation there is always a stretch of cones which do not allow any light to reach the microscope. Here all the light which passes the dioptrics emerges between the bases of the cones, i.e. where the iris tapetum lies. We have already seen that the space between the pseudopupils is the colour of this pigment for this reason. From the foregoing it follows that, within certain limits, the deflection undergone by rays at the boundary surfaces between two facets is greater the further the facet is from the meridian plane, and also, on the meridian, the further the facet is from the axis of the phenomenon. It is no wonder that the accessory images are not more distinct, rather one is amazed that they are as clear as they are. With this explanation of the accessory images, the origin of the first order accessory pseudopupils is also accounted for, since we need only to imagine the ray path in reverse, and to consider that all the rays which compose the accessory image will emerge in the pigmented cone margin, and it is clear that six, separated black spots must be seen around the principal pseudopupil.

I am well aware that, in many respects, the account given here is deficient, for example I have not taken into account the curvature of the corneal facets, nevertheless I believe that the gist of this explanation for the complicated phenomenon is correct.[64]

The proposed theory predicts that no animal with opaque pigment between the corneal facets (e.g. Fig. 60) should show accessory pseudopupils, and indeed I have never seen accessory pseudopupils in such an insect. The same holds for the accessory images. The firefly and the relatives of *Cetonia* show accessory images but no accessory pseudopupils, the latter for the same reason that they have no principal pseudopupil. Both principal and accessory pseudopupils are present, but cannot be distinguished from the intervening space because these animals only have black pigment in their eyes. Similarly, I can now explain why the great green bush cricket (*Locusta viridissima*) has no accessory pseudopupils despite having such brightly coloured eyes. Histological sections showed that the light-coloured pigment in these animals reaches almost to the tips of the cones. Under these conditions the accessory pseudopupil obviously cannot not stand out from its surroundings.

This pigment distribution is also the reason why the principal pseudopupil of this animal is so exceptionally small. All else being equal, it is simple to see that the principal pseudopupil must be larger the further the iris pigment extends up the cone or the less deep the iris tapetum reaches down.

Everything I have said about the relative position of the corneal reflection and the luminous pseudopupil is also valid for the entire pseudopupil phenomenon.

Chapter X Vision with Compound Eyes

A. Resolution of the Retinal Image

I am only able to make a definitive statement about the resolution of the dioptric retinal image for the case of *Lampyris*. The resolution is indicated in the frontispiece, which shows that, as far as the retinal image is concerned, the animal would still be able to read sign writing at a distance of some metres. The main stroke of the R on the window was 4.9 cm wide, and the distance from the eye 2.25 metres. Although the photography and further reproduction can certainly only have degraded the image, the "R" is still recognizable. In ophthalmological terms the retinal image of *Lampyris* would correspond to a visual resolution of 6/400 to 6/500 Snellen.[65]

The firefly would thus still be able to resolve a grating with 4.9 cm wide bars at a distance of 225 cm. It would also receive a retinal image of the same size when the bars and the distance are proportionally reduced. Thus at a distance of 1cm it would still resolve a grating with bars of 0.22 mm. As one sees, the eye's performance cannot be called poor and it should serve the animal very well in searching for food etc.

From the structure of their eyes it can hardly be doubted that other insects and crustaceans with superposition eyes have much sharper retinal images; their size alone is strong indication of this. With a little patience I believe it would be possible to render these retinal images directly visible, as I have done in *Lampyris* and relatives, or to gain a measure of their resolution from ophthalmological data.[66]

The same is true for apposition eyes. Thus, for example, in *Limulus* one could derive a theoretical value for the resolution from the data provided (Chap. III), if the diameter of individual retinulae in the living animal (or in sections of a frozen eye) were to be determined.

One can ask whether the resolution of the retinal image is really a measure of visual acuity. In theory the retina might be incapable of resolving the details in the image. However, this suggestion can hardly be taken seriously. Firstly the number of retinal elements, which is the same as the number of facets (if, like Grenacher, we consider the rhabdom as a physiological unit), corresponds to the resolution of the image. Secondly it would be absurd to think that nature had provided the animal with such a complex dioptric apparatus without a retina to analyze its image. Admittedly, it should not be claimed that the image is perceived in its full detail; below I will describe a situation which makes this seem questionable.

B. Distortions of the Retinal Image

Worthy of special consideration is the fact that frequently, perhaps indeed in the majority of species, the retinal image of the compound eye is geometrically dissimilar to the projection of the object. Amongst the vertebrates such distortions are of minimal significance.[67] However, in compound eyes we have seen arrangements where such a distortion of the retinal image serves to broaden the visual field, i.e. the obliquely inclined cones at the eye margin. We also often saw very considerable deviations of the corneal surface from a spherical form, i.e. very marked differences in the radius of curvature along different meridians, which then naturally also lead to distortions of the pseudopupil (dragonflies, *Carcinus*). The greatest deviation was seen in *Squilla*, and the consequences of this were discussed previously.*

In general, when the corneal surface is not in the form of a spherical shell around a spherically curved retina, or when, with a spherical surface, the ommatidial axes are not radially arranged, then a distortion of the retinal image must always result whereby its geometrical similarity with the visual field is lost.

For example, the stronger curvature of the eye in the periphery has the consequence that the retinal image of a square is compressed vertically in the dorsal and ventral retina and horizontally in the lateral retina. In intermediate positions, rhomboids result.

An analogous situation must occur in the centre of the retina, when its radius of curvature differs along different meridians. For example if it is longer in the horizontal direction, a square in the middle of the visual field will be imaged as a rectangle with a longer horizontal side.

Can this geometrical dissimilarity of the retinal image to the visual field be considered detrimental for vision?

From a physiological point of view such impairment is not to be expected. Thus we recognize the size of an object rather well whether it is close or far, i.e. whether its retinal image is large or small. Unless he has learnt it in school, a man is not aware that he has a retinal image or how it is shaped; an arthropod even less so. The value of all sense organs in the perception of the outside world is rather based on the fact that under identical external conditions, identical nervous excitation is transmitted centrally. Under constant conditions of the animal and its organs, a change in the outside world is recognized from differences in the information reaching the brain. The retinal image in a given eye is always distorted in a particular way. The insect in question has always seen a bird flying by at the edge of the visual field with an elongated retinal image, and if now such an elongated retinal image occurs in the same part of the visual field the insect, unaware of retinal images, will recognize the bird and try to conceal itself.

Naturally, it cannot be said that this individual has learnt to interpret its retinal image, but presumably the entire central mechanisms of instincts etc. have developed over the generations on the basis of this particularly shaped retinal image.

---

* Müller (1826) has already noticed and discussed the deviation of the eye from a sphere, but he points out that the clarity of the retinal image differs in different positions as a result.

A retinal image, distorted as described, is thus by no means disadvantageous for the survival of the individual. On the contrary, its form is presumably adapted to the needs of the organism. Even though less details are contained in the retinal image at the periphery, the broadening of the visual field may be of greater importance for the animal.Thus if it wishes to see details on an object it can always direct the centre of its eye towards it. We do exactly the same. The centre of our retina serves for exact observation: we direct our gaze at objects which intitially aroused our interest in peripheral parts of the visual field.

C. Seeing Movements

In the foregoing I have intentionally only discussed the resolution of the retinal image and its distortion. Just as the distortion of the retinal image does not result in the perception of a distorted object, the resolution of the retinal image cannot be taken directly as a measure of the animal's resolution. It is impossible for the animal to see details not contained in the retinal image, but as already remarked, it is quite possible that details are missed which are present in the retinal image. Here the performance of the retina and the whole nature of its function are important.

In general, the retina in the compound eye is thicker than that in the vertebrate eye.[68] Considering the small dimensions of the dioptric apparatus the relative difference is thus immense. If only a thin layer of this retina were light-sensitive, then the image could be perceived in full detail. However, nowhere is a thin layer, which could fulfil this function, to be found. In fact we are probably right in assuming that the retinulae or rhabdoms serve this function. But the retinula layer is very thick, and in many eyes can always be reached by light not entering the facet exactly on axis. In other eyes this is only true when the retinal pigment is in the dark-adapted position, and in some (e.g. butterflies) probably not true at all or only to a very small extent. Thereby, light from a bright point inevitably excites not only one retinula, but also – though to a lesser extent – the neighbouring ones. A glance at Text Fig. 11 (p. 27) shows that when the image point *B*, lies, for example, at the distal end of a retinula in the *Lampyris* eye, the proximally diverging rays must also penetrate neighbouring retinulae. The manner in which the rhabdoms retain the captured light will certainly greatly restrict this diffusion of light, but probably cannot eliminate it altogether. Thus a bright point, even though its retinal image observed with the microscope is sharp, will still be perceived as being surrounded by a halo whose intensity decreases rapidly from the centre. If the image point were to lie, not at the distal end of the rhabdom, but at the proximal end or somewhere in the middle, this would not make any fundamental change to the behaviour of this light spread.

I believe that the physiological significance of the light-induced pigment migration in crustaceans can be understood alone in these terms. We have already seen how the perceived retinal image gains relatively in brightness at the expense of resolution during the night. This seems particularly desirable as the retinal image in the compound eye, generally speaking, is much less intense than that in the vertebrate eye. Compare, for example, the large pupil forming the base of the ray cone for each image point in mammals, with the small group of facets which form this base in the superposition eye, or the single facet in apposition eyes.

The resolution of the retinal image thus gives only the upper limit for visual resolution. The physiological blur reduces this, though not to such an extent that one cannot make a reasonable estimate of visual resolution from the resolution of the retinal image. The above leads me to the question: what is the significance of the two types of eye which occur in the animal kingdom? Do they really have identical functions, and has nature achieved the same goal via two fundamentally different paths, once with one lens system and an inverted image, and once with hundreds of lens systems and an upright image? If yes, then why do many animals have small simple eyes next to their large compound eyes? What is the point of the small ones if the large ones perform exactly the same function, only better because of their larger size?[69]

In the answer to this question I still hold the same view which I stated in my first work on the compound eye (Exner 1875). Admittedly the exact justification for this must now be modified in the light of my present day knowledge, but this only strengthens the theory.

In my opinion the vertebrate type eye is better suited to the recognition of shapes of external objects, and the compound eye better suited to the recognition of changes in objects.*

The perception of changes, especially movements of objects, plays a much greater role than is usually assumed. This is correlated with enemies from which one must protect oneself and with prey which must be hunted. Again and again I have convinced myself that deer or hares do not recognize humans by their shape; if one remains perfectly still one seems to be taken for something strange and they watch one long and attentively, but they then come up very close and finally become completely at ease with the foreign object. Anyone who has closely observed a cat knows how the slightest movement in their visual field will arouse their attention, whilst they are relatively blind to stationary objects. An adder does not see a frog in its cage whilst it is still, but with the first jump it is seen and caught. Any angler knows that his fly must move on the water surface if the trout is to bite. A great number of such examples can be cited amongst the vertebrates; for insects movement appears even more important. For example, when trying to catch a butterfly or a dragonfly, any sudden movement startles the animal, but with a sufficiently slow and careful approach one can often succeed. The shape of the moving object appears rather irrelevant, it flies away whether a bird flies past, a butterfly net is carelessly brought near, a man or a dog passes by etc. The insect knows its enemies not by their form but only by movements. A fly sits fearlessly on any stuffed bird, but a whole swarm of flies takes off when a handkerchief is waved in the room. Conversely the dragonfly recognizes its prey mainly from movements, since it dives onto a paper ball thrown into the air just as it does onto flying insects.

* Earlier I wrote "of movements in objects". It appears justified to broaden the concept to "changes" following studies on the ability of the human eye to perceive motion and changes similarly.

Before I began my studies on compound eyes I had come to the view, on the basis of studies on physiological optics, that the perception of movements is a basic sensation comparable to the distinction between dark and light or red and another colour. As for the sensation of colour, the sensation of movement also has an upper and lower intensity limit and it also leaves behind a negative after-image which is again restricted to the stimulated region of the retina etc.

Of importance to us is the fact that the sensitivity to movement at different places in the human retina is by no means correlated with the ability to localize objects. The latter is best in the centre and reduces rapidly towards the periphery. In contrast, the reduction in sensitivity to movement towards the periphery is so much less marked that one is inclined to say that the major function of the peripheral retina is the perception of movement, or as I later concluded, the perception of changes in general.* In fact, daily experience shows that we never notice stationary objects unless they lie very near to the point of fixation, but the slightest change in the peripheral field arouses our attention immediately, and make us direct our glance almost as a reflex action. von Fleischl (1883) has proposed an ingenious hypothesis about the structural organization of the human retina which can explain this correlation between relative hypersensitivity to movement with reduced spatial sensitivity and which ties in with my views on the functioning of the insect eye.[70]

Namely, I take the view that the compound eye functions in a similar way to the human peripheral retina, i.e. that the perception of changes is of prime importance, whilst the perception of forms is secondary.[71]

The reason for this view comes from the following consideration: as just discussed, the superposition image of a single bright point, even if the optical image were perfect, excites not only one retinula but a whole group. A small blur circle of sensation arises, within which, however, excitation is not evenly distributed but reduces continuously to zero away from the centre. Let us now consider the bright point moved so little that its retinal image moves only the diameter of one retinula. Then the degree of excitation in all the retinulae within the blur circle must have changed. It is clear that this change in excitation in a large number of nerve endings is extremely well suited to attract attention, i.e. to detect the movement as well as its direction. The same is true for any change, e.g. the sudden appearance of a previously invisible object (this is possible without a very noticeable movement).

The blur circle in a properly constructed vertebrate eye would not work in the same way because it is of equal intensity.** In the same situation changes in excitation occur only in the relatively small number of retinal elements forming the margin of the blur circle. The two types of blur circles thus behave very differently, and the physiological effect of the movement of blur circles of the same size should have the relationship of circumference to area. Thus in this

* For example, when a group of dark points on a bright background is moved so far into the periphery of the visual field that one can no longer count them even approximately, one nevertheless notices immediately when one disappears or another appears.

** I say a properly constructed eye, since in fact brightness differences do occur, which, however, are irregular and thus could not function like the blur circle in compound eyes described above.

respect the compound eye has a very significant advantage. Of course, a comparison of the movement sensitivity in the two eye types assumes that in both cases the retina is similarly sensitive to changes in intensity. Naturally any estimate of an eye's performance from its dioptrics is based on this assumption.

As argued below, it is probable that apposition eyes also have a blur circle of this sort.

One recalls that in apposition eyes an image of a small part of the visual field is formed in the vicinity of the cone tip (see Text Fig. 9, p. 17, *ab*). In the corresponding retinula, however, it will not be perceived as an image, but only as a unitary signal, since each retinula represents only one visual element. Like the cone margin, the outer surface of the retinula is surrounded by pigment. How much of the image *ab* falls on the retinula depends upon the position of the retinula and the surrounding pigment. It would be very inefficient (and hence unlikely) if the part of the image perceived was smaller than the corresponding part of the visual field. Thus if it were smaller the animal would not receive signals from all parts of the visual field, but the visual field would, so to speak, be covered in a net, the animal seeing through the meshes which conceal much of the view. It would have no more sense of the presence of this net than we do of our blind spot, nevertheless it would experience a significant impairment to its vision. One would thus assume that the perceived parts of the image, *ab*, in neighbouring ommatidia would be at least large enough for the corresponding parts of the visual field to touch each other. One now only needs to assume that they are a little larger, and then this eye type also has blur circles which behave just as in the superposition eye. Thus in this case the retinula of one ommatidium receives not only light coming from the corresponding part of the visual field but also light from points in the field belonging to the neighbouring ommatidia, but with intensity diminishing with the distance from the axis.

Actually, from the observations on eye cup preparations (Chap. V) this assumption is not even necessary and it is not a hypothesis, but a fact that the inverted image of the same bright object is seen at the tips of a whole (albeit small) group of cones; and from the anatomy of the retinulae it appears hardly possible that only light from the geometrically corresponding part of the visual field reaches each one. If the observations of Leydig and Thompson Lowne are true, namely that the pigment diaphragm between the cone and retinula can change its diameter in vivo (admittedly this possibility can only be considered for certain species), then we also have here, as with superposition eyes, a mechanism whereby not only the brightness but also the size of the blur circle, and thereby the resolution of the image, can be varied.

I place a greater emphasis on the recognition of changes in insect vision, because it is here that I see the most fundamental functional difference from vision with the inverted retinal image; however, by no means do I underestimate the recognition of forms.* The elegant experiments of Plateau (1887/1888)[72] have adequately established that insects see objects and their position, and the resolution of this performance agrees very well with my own suppositions.

* Cf. the discussion of the *Squilla* eye.

Incidentally it seems obvious to me that amongst different species the relationship between the development of form and motion detection is subject to significant variations.

A butterfly's behaviour in flying from flower to flower indicates that it sees these from a moderate distance. The fact that it is not too infrequently deceived by objects of similar shape and colour, first noticing the deception when it is almost in contact with the object, indicates that its vision is developed to approximately the level suggested by the retinal image of *Lampyris*.

One can also readily deduce that the animals have colour vision, principally from the way in which they can be deceived; this is well-known from the exact experiments of Lubbock,[73] and the whole colourful world of butterflies and flowers is witness to the fact.

D. Accommodation

I find it very improbable that accommodation occurs in compound eyes. Firstly, this is superfluous in apposition eyes since a small shift in the image projected onto the cone tip is hardly of physiological importance. Secondly, neither in apposition nor in superposition eyes can one find arrangements which could conceivably function as an accommodation mechanism.[74] In superposition eyes, on approach of an object the image shifts distally, in contrast to vertebrate eyes. In *Lampyris* I estimated a shift of ca. 0.092 mm when the object was brought from a distance of 810 mm to 1.2 mm. In order to bring the object so close I projected an image of the object through an Abbe illuminator. One can hardly doubt that the male firefly uses its eye to locate glowing females in the grass, an object effectively at infinity. Whether he still needs to see clear images at a distance of 1.2 mm would appear questionable.

If an accommodation mechanism were present it would have to bring the retina and dioptrics closer together on fixation of nearby objects. But there is no trace of such a mechanism. It is also easy to see why no accommodation mechanism is required in this eye. Thus the large thickness of the retina can completely substitute for it. If the image moves somewhat back or forth it can still remain within the retina. In the vertebrate eye with a similar movement of the object, the image has moved far behind or in front of the retina.

Another consideration is also valid for vertebrate eyes. Assuming that the sensitive layer of the retina has the same absolute thickness, the need for accommodation decreases with the size of the eye. Thus, all else being equal, the shift in the retinal image is proportional to the linear dimensions of the eye – for which the back focal length of the lens is a measure.* In elephants, for example, this is many times greater than in the mouse; and since sensitive retinal layers and the retinal elements show only small size differences (in frogs they are even greater than in elephants), at a distance where the elephant must already show strong

* The distance of the image from the second focal point $f_2 - F_2 = D$. Since $f_2 = F_2 f_1 / f_1 - F_1$, $D = F_1 F_2 / f_1 - F_1$

accommodation, the mouse still receives a sharp retinal image. All the same, whereas objects viewed at a distance of several millimetres are quite insignificant for the elephant, they could still be relevant for the mouse.

The situation is very similar for insects and crustaceans with superposition images, the eyes of which are of the same size as those of smaller mammals.

E. Depth Vision

It can hardly be doubted that insects and crustaceans can distinguish close and distant objects. Particularly in insects, simple experiments show that they are more frightened of a close object even when apparent size and speed of a more distant object are greater. Humans, and apparently all vertebrates, recognize distances mainly on the grounds of the difference in the retinal image of the same object between the two eyes, and from the degree of convergence required to fixate an object binocularly. The latter mechanism does not come into consideration in many animals with compound eyes, since their eyes are fixed immovably on the head. In other species (many crustaceans) the eyes are mobile, but their movements are so independent of each other that it is difficult to believe that a mechanism of binocular fixation plays a major role. Further, the position of the eyes on the head is often such as to preclude this possibility.[75]

However, there is no reason to assume that the first mechanism does not play a similar role in compound eyes as it does in vertebrate eyes. Here, e.g. in dragonflies, there is a large binocular overlap, undoubtedly much larger than in humans, extending not only frontally but also dorsally, ventrally and posteriorly. Here as well, the images in the two eyes must be different, and the two retinal locations receiving the image of the same object point will determine its distance and position uniquely for a given position of the head.[76]

The accuracy of binocular depth perception in vertebrates increases with the distance between the two eyes (Berlin 1889). In insects this distance is, of course, very small, although it tends to be as large as the dimensions of the head allow.[77] It does not follow that this kind of depth perception is beyond them, but only that it is less perfect at large distances. At close quarters this principle of depth perception can function exactly as in vertebrates. A vertebrate has no more useful depth perception a few centimetres away from the eye due to the poor quality of the retinal image and the impossibility of further convergence. Perhaps it is at just this distance that a dragonfly has its most accurate depth perception, and this may be more advantageous than an accurate estimation of greater distances.

It is quite possible that this is why some insects, as observed by Plateau, change their direction first only shortly before an obstacle. They may have seen the object before, but only first recognized its distance, and hence its nature as an obstacle, when they were close enough to obtain a useful binocular impression of depth. The reason why many insects first hit a sheet of paper before changing direction may be that depth perception is only possible when details of the object can be seen (as in vertebrates as well). Thus the distance of the paper, if evenly coloured, could only be judged by perception of its edges.

As I said, because of the independent movements of their eye stalks, it is difficult to conceive of binocular depth vision in many crustaceans. An alternative mechanism is suggested by their lively eye movements. To aid in the recognition of difficult objects we commonly stretch out our necks and move our heads to and fro. Animals also do this on occasion. The reason is to observe the object from slightly different perspectives. The successive images thus received guarantee a more correct view not only of the form but, above all, of the distances of the different components. von Helmholtz (1867) gives a very pertinent example whereby one can convince oneself of the value of successive images: if one closes one eye and then, with a steady head, looks at the twigs and branches in a tree or bush one sees only a confused jumble of twigs and leaves, and above all one can gain no impression of the relative distances of individual branches. But as soon as one moves one's head to and fro, the whole scene crystallizes spatially, boughs and branches separate, in short one obtains a three-dimensional image. Naturally this is due to images of closer objects moving more rapidly across the retina than those of distant ones.

I thus consider it quite likely that crustaceans, whose eyes do not make coordinated movements, but move to and fro almost playfully, obtain their perception of depth in this way. This is, of course, only an hypothesis, the verification of which is unfortunately not easy but may not be insuperable. It also receives support since these almost continuous movements have no other obvious function.

Chapter XI Some Remarks on the Phylogeny of Compound Eyes Considered from a Functional Viewpoint[78]

Grenacher (1879) recently discussed the phylogenetic relationship between compound eyes and the simple stemmata, the latter built on the plan of the vertebrate eye.

It is indeed true that in related animals, and even on one and the same head, one finds the two eye types side by side, one seeing with an inverted, the other with an upright retinal image. What conceivable link can there be between the inverted image of a camera obscura and the upright superposition image formed by many astronomical telescopes?

In conclusion I want to make a few remarks to this end from an optical viewpoint, particularly as my considerations lead to basically the same conclusions as Grenacher reached on morphological grounds. As a rough approximation, a nipple-shaped swelling of the cuticle, combined with one or a few underlying sense cells capable of converting movements of the ether into nervous excitation, perhaps via heat production in black pigment, can be considered as the primitive eye. Such primitive eyes can still be seen today in lower animals. For a given position of the animal they are capable of distinguishing light, dark and direction. With such an eye, an animal swimming in water would be able to determine the direction of the surface.

There are now two possibilities for the development of this most simple visual apparatus. First the cuticular thickening (the human lens is also of ectodermal origin) takes on more and more the shape of a lens, and the sense cells come to lie near the focal plane. Even with the most simple nipple-like thickening the sense cells lying on the axis will in general receive the most intense excitation when the axis is directed at the sun. Thus a primitive ability to localize would already be present when one assumes several sense cells in the primitive eye. If we imagine these cells as gradually increasing in number and migrating towards the focal plane, the cones of light from the sun would be absorbed more and more intensely by the sense cells, so that the ability to localize would steadily improve. Thus, without a jump – this is after all the important point – the steadily increasing value of the organ leads to the simple but highly developed eyes of spiders etc., or vertebrates. The sense cells spread out in large numbers in the focal plane of the refractive media, i.e. on a concave surface, and receive the inverted image.

Secondly, the primitive eye can develop, via multiplication, into the compound eye. At a given moment an animal with a primitive eye can only tell if there is a bright object in the direction of the axis and its surround. To learn about other directions it must move its body. To gain information from several directions simultaneously, several eyes with divergent axes would have to develop. Two such eyes working together already provide the beginnings of an upright image. The greater the number of these primitive eyes the more perfect the localization. Thus, again without a jump, the numerous radially arranged ommatidia of the compound eye can arise, each one a relict of the original element. Because of the small size of the cuticular lenses, the light penetrating individual eyes will be small. Thus structural developments, be they the formation of spherical refracting surfaces or the layering of differently refracting media, will arise, which guide as much as possible of this meagre light to the nervous element(s). In this way we come directly to the apposition eye.

In morphological terms we have seen this developmental series in the foregoing, starting with the trilobites and progressing through *Limulus* to the diurnal insects. In this series one sees clearly how the inverted image of the single eye (in trilobites) is lost, allowing for the perfect development of the upright image of the whole eye (also already represented in the aggregation of eyes in the trilobites).

The question now arises of whether there is a phylogenetic bridge from the apposition to the superposition eye, i.e. if a continuous series of steps is conceivable, in which each confers an advantage over the former in the struggle for existence, thus allowing further development. It seems to me that this continuous series is not hard to find.

The dioptrics of the apposition eye includes, as we saw, a lens cylinder of approximately the same length as its focal length. It functions so that an image of a narrow portion of the visual field is formed at the cone tip. This results in a relatively intense light stimulus, because all the principal rays from the different object points in the visual field enter the retinulae more or less parallel to the cone axis. Thereby the ray cones from each image point show a not inconsiderable divergence. Let us consider an animal with such an eye forced to become nocturnal or crepuscular for biological reasons. Then in order to see better in dim light the cones need only to become longer and the retina to move correspondingly further away. Thus through either lengthening or increasing the power of the lens cylinder the divergence of the ray cone belonging to each object point is reduced, which means an increase in the perceived intensity for the centre of the "elementary visual field" (see p. 20). At the same time the principal rays from peripheral points in the partial visual fields no longer tend to run parallel to the ommatidial axis. They form an angle with it in the same sense as in a superposition image. Thus, if the retina has receded, a part of the ray cone from this peripheral point now hits the neighbouring retinal element, and so is no longer lost for perception. On pp. 58 and 61 I have already mentioned such eyes, which represent transitional forms between the two types.

In this way the relative brightness of the retinal image can gradually increase to the advantage of the animal in question without a perfect superposition image being formed. This first arises, as discussed above, when the original lens cylinder has doubled in length (or power) i.e. having twice the length of its focal length, and the retina has receded very considerably.

Thus in dim light, it seems to me in all probability, this goal can be reached with continuous improvement of the eye's performance. It must also be remembered that the migration of the iris pigment can, at any time, convert the imperfect superposition image to an apposition image, should an excess of light demand it.

References Cited by Exner

In a very few instances the full reference could not be traced, in which case the partial information provided by Exner has been reproduced.

Berger E (1878) Untersuchungen über den Bau des Gehirns und der Retina der Arthropoden. Arb Zool Inst Univ Wien 1: 173–220

Berlin R (1889) Über die Schätzung der Entfernungen bei Thieren. (Festchrift zur Feier des XXV Regierungsjubiläums Sr Majestät des Königs Karl von Wurttemburg) Stuttgart

Boll F (1871) Beiträge zur physiologischen Optik. Arch Anat Physiol Wiss Med 1: 530–549

Brehm AE (1878) Brehms Thierleben. Leipzig

Brücke E (1852) Vergleichende Bemerkungen über Farben und Farben-wechsel bei den Cephalopoden und bei den Chamaleonen. Sitz Ber Akad Wiss Wien 8: 196–200

Brücke E (1887) Vorlesungen über Physiologie. vol 2. Wien

Ciaccio GV (1889) Sur les facettes de la cornée et les milieux réfringents des yeux composés des Muscidés. J Microgr 13: 80–84

Claus C (1879) Der Organismus der Phronomiden. Arb Zool Inst Univ Wien 2: 59–128

Exner K (1886) Zur Linsenformel. Linsenwirkung nicht homogener Körper. Ann Phys Chem 28: 111–116

Exner S (1875) Über das Sehen von Bewegungen und der Theorie des zusammengesetzten Auges. Sitz Ber Akad Wiss Wien 72: 156–191

Exner S (1885) Ein Mikrofractometer. Arch Mikr Anat 25: 97–112

Exner S (1886a) Über Cylinder, welche optische Bilder entwerfen. Pflüger's Arch Ges Physiol 38: 274–290

Exner S (1886b) Nachtrag zu der Abhandlung Über "Cylinder, welche optische Bilder entwerfen". Pflüger's Arch Ges Physiol 39: 244–245

Exner S (1889a) Tagblatt der 61. Vers Dtsch Naturforsch Ärtzte, Köln, p 71

Exner S (1889b) Das Netzhautbild des Insectenauges – Durch Licht bedingte Verschiebung des Pigmentes im Insectenauge, und deren physiologische Bedeutung. Sitz Ber Akad Wiss Wien 98 (3): 13–65

Fick E (1889) Über Lichtwirkung auf die Netzhaut des Frosches. Ber Ophthalmol Ges Heidelberg – –

Fleischl E von (1883) Die Vertheilung der Sehnervenfasern über die Zapfen der menschlichen Retina. Sitz Ber Akad Wiss Wien 87 (3): 246–252

Forel A (1886) Sensation des Insectes. Recueil Zool Suisse 4: 1–50

Forel A (1887) Sensation des Insectes. Recueil Zool Suisse 4: 145–240

Girschner E (1887) Einiges über die Färbung des Dipterenauges. Berlin Entomol Z 31: 155–162

Gottsche A (1852) Beitrag zur Anatomie und Physiologie des Auges der Krebse und Fliegen. Arch Anat Physiol Wiss Med 483–492

Grenacher H (1877) Untersuchungen über das Arthropoden-Auge. Beilagenh Klin Mon Bl Augenheilkd 15: 1–42

Grenacher H (1879) Untersuchungen über das Sehorgan der Arthropoden: ins besondere der Spinnen: Insecten und Crustaceen. Vandenhoek & Ruprecht, Göttingen

Gruel C (1844) Mikroskopische Beobachtungen. Ann Phys Chem 61: 220–222.

Helmholtz H von (1867) Handbuch der physiologischen Optik. Voss, Leipzig

Klemensiewicz R (1878) Beiträge zur Kenntniss der Farbenwechsel der Cephalopoden. Sitz Ber Akad Wiss Wien 78: 7–50.

Kühne W (1877) Eine Beobachtung über das Leuchten der Insectenaugen. Untersuch Phys Inst Univ Heidelberg 1: 242–247
Leydig F (1855) Zum feineren Bau der Arthropoden. Müller's Arch. Anat Physiol 22: 406–444
Leydig F (1857) Lehrbuch der Histologie des Menschen und der Thiere. Meidinger, Frankfurt
Leydig F (1864a) Das Auge der Gliederthiere. Laupp, Tübingen
Leydig F (1864b) Tafeln zur vergleichenden Anatomie. Tübingen.
Matthiesen L (18–) Centralbl Opt Mech 7. 10
Matthiesen L (1886) Über den Strahlendurchgang durch coaxial continuirlich geschichtete Cylinder mit Beziehung auf den physikalisch-optischen Bau der Augen verschiedener Insecte. Repert Phys 22: 333–353
Müller J (1826) Zur vergleichenden Physiologie des Gesichtsinnes des Menschen und der Thiere nebst einem Versuch über die Bewegungen der Augen und über den menschlichen Blick. Cnobloch, Leipzig
Müller J (1875) Kosmische Physik, 4th edn.
Notthaft J (1880) Über die Gesichtswahrnehmung vermittelst des Facettenauges. Abh Senckenberg Naturforsch Ges 12: 35–124
Patten W (1887) Studies on the eyes of arthropods. J Morphol 1: 194–226
Plateau F (1887) Recherches expérimentales sur la vision chez les Arthropodes. Bull Acad R Soc Belg Ser 3, – –
Plateau F (1888) Recherches expérimentales sur la vision chez les Arthropodes (3ième partie). Bull Acad R Soc Belg Ser 3, 15: 28–91
Schmidt O (1878) Die Form der Krystallkegel im Arthropodenauge. Z Wiss Zool 30: 1–12
Schultze M (1868) Untersuchungen über die zusammengesetzten Augen der Krebse und Insecten. Cohen, Bonn
Sharp (1889) Address read before the Entomological Society of London (16 Jan)
Thompson Lowne B (1884) On the compound vision and the morphology of the eye in insects. Trans Linn Soc London Ser 2 Zool 2: 389–420
Watase S (1890) On the morphology of the compound eye of Arthropods. Studies Biol Lab Johns Hopkins Univ 4: 287–334

Plates

Plate I

Fig. 1. Meridional cross-section through lateral part of the eye of *Lampyris splendidula.* Fixed in alcohol, embedded in celloidin; stained with saffron. Magnification 120 ×. *c.* corneal facet *I.P.* Iris pigment *K.* The tips of the crystalline cones. The shape of a cone and a facet are shown in Text Fig. 12 p. 30 *G.* Vitreous body of other authors, filled with spindle-shaped radially arranged cells. *R.* retina *RP.* Retinal pigment *m.f.* Membrana fenestrata, penetrated proximally by nerve bundles leaving the *G.o.* Optic ganglion.

Figs. 2–5. Optical cross-sections of the ray bundles emerging from the crystalline cones of the *Lampyris* eye, seen under low magnification when using a single light source as object. The ray path and the four levels of focus are shown in Text Fig. 11 p. 27. **Figs. 2, 3, 4, 5** correspond to the levels AA_1, CC_1, DD_1, EE_1 respectively. **Fig. 2** shows the rays (or their projection) when focused at a level within or distal to the crystalline cones. Focusing towards the retina the individual ray bundles come closer together giving the picture seen in **Fig. 3** at the level of the crystalline cone tips, CC_1. In fresh preparations (these figures come from a fixed animal) the haloes are less pronounced. Focusing further back, the ray bundles converge into a small bright spot at the level of the retina DD_1. Even further proximally they diverge into a blurred circle (**Fig. 5**) showing elegant diffraction patterns.

Figs. 6, 7. Crystalline cone and distal retinula in *Dytiscus marginalis.* Drawn in Camera lucida × 550. **Fig. 6** shows the iris pigment in the light-adapted position. **Fig. 7** dark-adapted.

Fig. 8. A portion of the eye of a living *Limulus. A* is in the region of the black pseudopupil and *B* outside it, so that the diagram shows the light distribution in the individual facets. Direction of illumination was from the right-hand side of the observer.

Fig. 9. Corneal facet and attached cone from *Cantharis fusca. A* killed in the dark and *B* in sunlight.

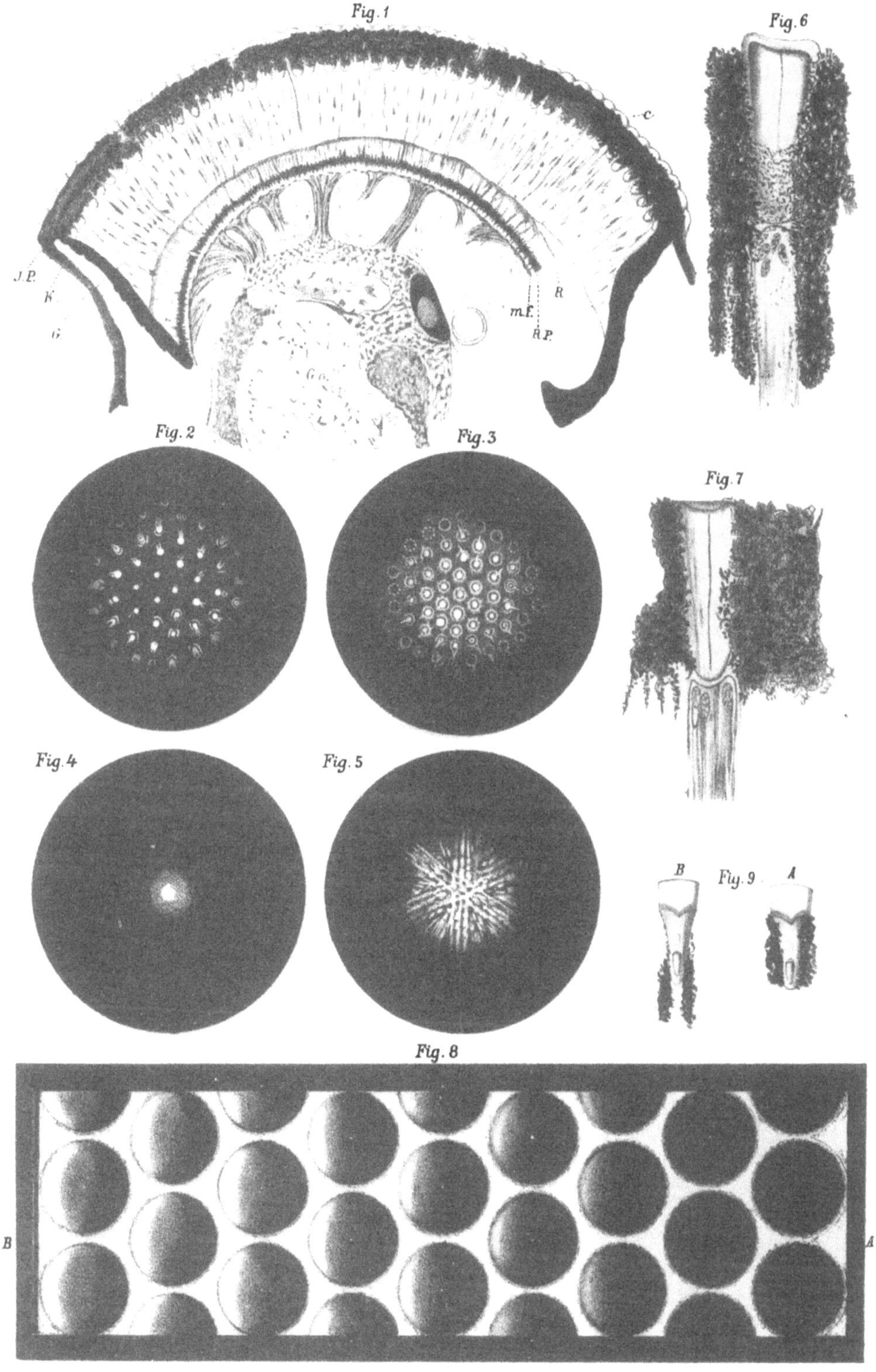
Fig. 1
c
J.P.
K
G
m.f.
R
R.P.
Fig. 6
Fig. 2
Fig. 3
Fig. 7
Fig. 4
Fig. 5
B
Fig. 9
A
Fig. 8
B
A

Plate II

Figs. 10, 11. Both drawings are of the same freshly teased preparation of the noctuid (*Plusia chrysetis*) retina. It shows the tracheal bundles which form the tapetum in moths, and which surround the retinulae proximally. In transmitted light (**Fig. 10**) they appear dark, and in incident light (**Fig. 11**) bright. In **Fig. 10** one can see the retinal pigment behind the tapetum and the nerve fibres entering it. *S* the distal halves of the thick part of the retinula (indistinct in this fresh preparation).

Fig. 12. Two isolated tracheal bundles from the same animal. It can be seen how each originates from a small trachea.

Figs. 13, 14. Sections through the eye of *Hydrophilus piceus*. Cornea not shown. Camera lucida drawing ×210 *R.* retina *m.f.* Membrana fenestrata The vitreous body between the retina and the dioptrics is criss-crossed by fibres. The fibres are inserted on cone-shaped extensions of the iris pigment. **Fig. 13** dark-adapted, **Fig. 14** light-adapted.

Fig. 15. Ommatidia from the red underwing (*Catocala nupta*). Left dark-adapted, right in the light adapted position. ×216. At *c* a pigment layer tightly adjoining the cone has remained although the iris pigment has migrated proximally. It is the analogue of the iris tapetum. Like the iris tapetum in crustaceans it is reflective and is responsible for the brown colouration of the eye, against which the black pseudopupil can be seen in the daylight. *a.* cross-section through the swelling of the retinula. *b.* Retina according to Thompson-Lowne. *m.f.* Membrana fenestrata.

Fig. 16. Ommatidia from *Colymbetes fuscus*. ×160. *A* dark-, and *B* light-adapted. Both killed with alcohol. *i.p.* iris pigment *r.p.* retinal pigment.

Fig. 17. Ommatidia of humming-bird hawkmoth (*Makroglossa*). ×166. *a* comes from the eye shown in Fig. 63. This showed a fine green glow in the centre of the main pseudopupil, when investigated with the ophthalmoscope. The iris pigment is in the dark-adapted position. *b.* comes from another specimen, the iris pigment is approaching the light-adapted position. *r.p.* the retina pigment is arranged in long streaks between the retinulae *m.f.* Membrana fenestrata.

Fig. 18. Tangential ground section through the fossilized eye of *Phakops fecundus*. ×22.

Fig. 19. Meridional section of the same eye. ×22.

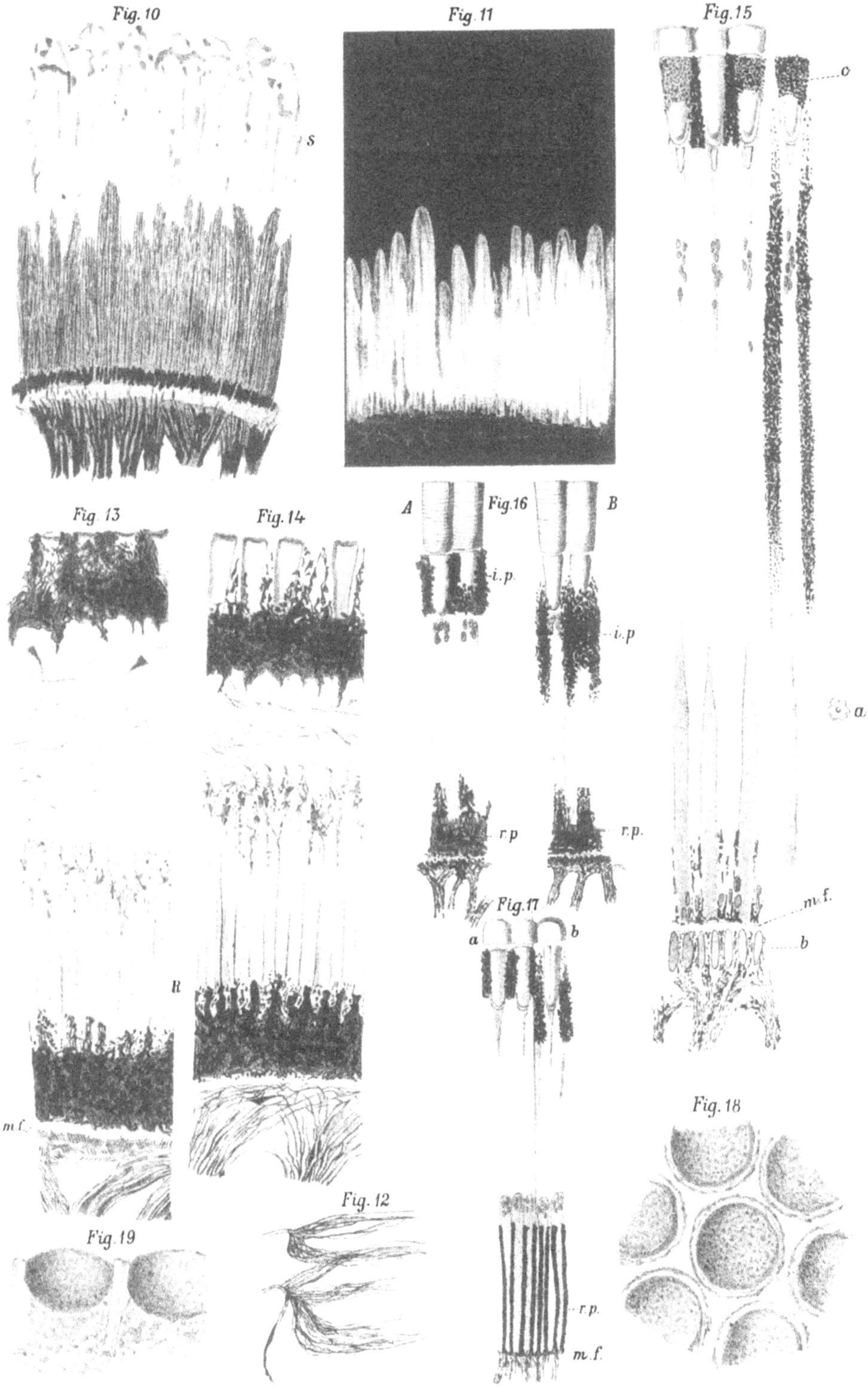
Fig. 10
s
Fig. 11
Fig. 15
o
Fig. 13
Fig. 14
A
Fig. 16
B
i.p.
i.p
a
r.p
r.p.
Fig. 17
a
b
m.f.
b
R
m.f.
Fig. 18
Fig. 19
Fig. 12
r.p.
m.f.

Plate III

Fig. 20. Ommatidium from *Limulus polyphemus* ×170. The pore canals in this ommatidium were not actually filled with air, so these have been drawn in from another ommatidium (*a*). The wave-like striation of the chitin is much richer and more regular than could be reproduced in the figure. *R* rods in the middle of the retinula.

Fig. 21. Cross-section through the retinula of the same animal. ×170. The radial structures in the middle are sections through the rods of Fig. 20. The section is somewhat more proximal to the level marked with *R* in Fig. 20.

Fig. 22. Section through the eye of *Squilla mantis* ×30. The section is parallel to the long axis of the cylindrically shaped eye and at its broadest point. *c.* cornea *k.* crystalline cone *I.P.* iris pigment *s.* retinulae *m.f.* Membrana fenestrata and retinal pigment in the vicinity *g.* optic ganglion.

Fig. 23. Three ommatidia from *Tropinota hirtella* ×280. *c.* cornea *T.* a crystalline cone completely ensheathed in iris pigment whilst both its neighbours are, at least distally, free. The retinulae contain granular button-shaped swellings adjoining the cones, and a prolonged thicker region proximally *m.f.* Membrana fenestrata *a.* indicates the level of the section shown in Fig. 24.

Fig. 24. The retinulae have been darkly stained. Between them are polygonal fields, probably cross-sections of tracheae; in long section (**Fig. 23**) they are seen as strongly outlined parallel streaks. ×1000.

Figs. 25, 26. Ommatidia of the bulldozer shrimp (*Scyllarus arctus*) ×85. **Fig. 25** dark-adapted, **Fig. 26** light-adapted. Both animals killed in alcohol and post-fixed in 0.1% Chromic acid. *a.* Cross-section of a retinula at the level indicated (higher magnification) *m.f.* Membrana fenestrata.

Fig. 27. Ommatidia from a hornet (*Vespa crabro*) ×132. The drawing is from the eye margin and shows the deviation of the ommatidial axes from the perpendicular to the cornea. The shading of the cornea and cones shows the appearance in the microrefractometer – otherwise the boundaries between the facets and the position of the cone are barely recognizable.

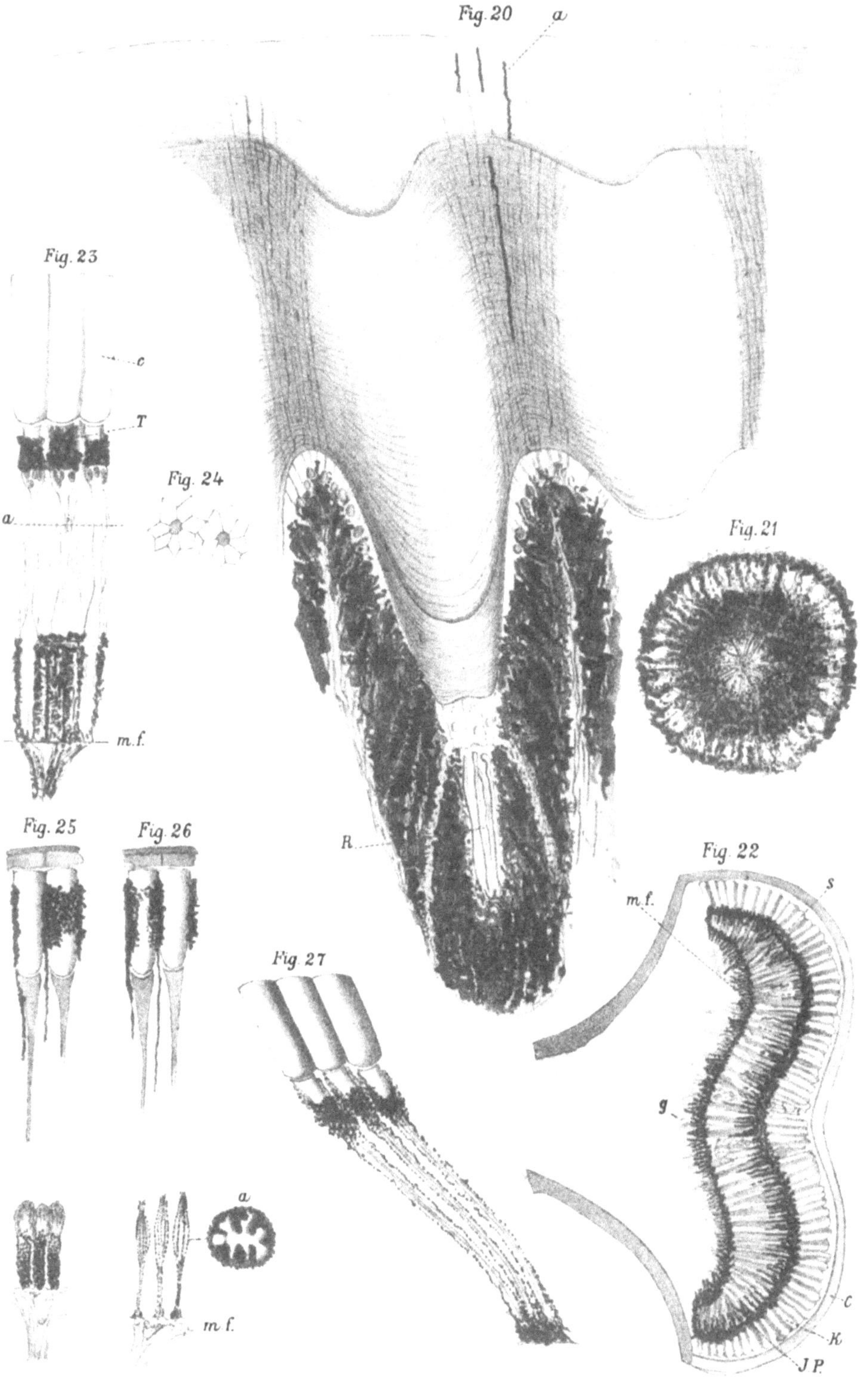
Fig. 20
a
Fig. 21
Fig. 22
m.f.
s
g
C
K
J P.
Fig. 23
c
T
a
m.f.
Fig. 24
Fig. 25
Fig. 26
R
Fig. 27
a
m.f.

Plate IV

Fig. 28. Ommatidia from the moth *Lasiocampa quercifolia*, × 145. The retinulae are composed of a thin distal and a thick proximal portion. *m.f.* Membrana fenestrata *n.* The retinal elements according to Thompson Lowne. *n.o.* Bundles of visual nerve fibres. The iris pigment is in the light-adapted position following one quarter hour in sunlight.

Fig. 29. Distal ommatidia from the other eye of the same animal in the dark-adapted state. Both figures show the muscle-like fibres attached to the pigment cells, which apparently contract on illumination.

Figs. 30, 31. Light- and dark-adapted eyes of a small moth (*Porthesia, Ocneria rubea, Leucoma salicis* or *Cnophria quadra*). × 208.

Fig. 32. Ommatidium of the butterfly *Epinephele* × 240. *m.f.* Membrana fenestrata *a.* club-shaped prolongation, presumably analogue of *n* (in Fig. 28) and *b* in Fig. 15.

Fig. 33. Ommatidium of the sawfly (*Tenthredo instabilis*) × 220.

Fig. 34. Two ommatidia of the cabbage white (*Pieris rapae*) × 180. A cross-section at the level *a* is shown in Fig. 35.

Fig. 35. The dark circles represent the retinulae, and the bright circles tracheae. The four rods within the retinulae are not fused into a common rhabdom. × 300.

Fig. 36. Two ommatidia of the small tortoiseshell (*Vanessa*) × 208. *a.* Part of a hair which sits on the cornea.

Fig. 37. Ommatidia from a dark-adapted *Portunus* × 125 *i.t.* Iris tapetum, which although dark is clearly distinguishable from *i.p.* the iris pigment. *r.p.* Retinal pigment following the nerve fibres out of the eye *m.f.* Membrana fenestrata – stretched thus allowing the connection between retinulae and nerves to be clearly seen *a. & b.* cross-sections of a retinula and pigment sheath at two different levels.

Fig. 38. Two ommatidia from *Portunus* kept in diffuse daylight. Otherwise as in Fig. 37. One can see the migration of the iris pigment and the retinal pigment – the latter having left the proximal ends of the retinulae and the nerve fibres almost pigment free.

Fig. 39. Shows two sections from the eye of *Sicyonia sculpta, D* in the dark-adapted and *L* in the light-adapted state. The third section, *A*, shows section *L* as it appears in incident light. The sections have been drawn together as if they came from one eye to emphasize the pigment layers. × 60. The individual ommatidia of this species are shown in more detail in Figs. 53, 54. Comparison of *L* and *A* shows that an iris tapetum (*i.t.*) lies on top of the iris pigment, and that the proximal ends of the retinulae are surrounded only by retinal pigment (*r.t.*). In the optic ganglion there is a third tapetal layer (*t.*) which is difficult to distinguish from pigment. Comparing the dark- and light-adapted eyes one sees that both iris and retinal pigment migrate en masse, the latter moving through the retinal tapetum up to the distal ends of the retinulae *C.* cornea *i.p.* Iris pigment *R.* retina *m.f.* Membrana fenestrata *i.t.* iris tapetum *r.t.* retinal tapetum. The light-adapted eye was from an animal kept in bright diffuse daylight.

Fig. 40. Ommatidia of *Lycaena* (chalk-hill blue) × 330. *m.f.* Membrana fenestrata with underlying nuclear structures as in many other lepidopterans.

Figs. 41, 42. Ommatidia from *Maja* × 126. **Fig. 41** Light-adapted, **Fig. 42** dark-adapted. The migration of retinal pigment is more conspicuous than that of the iris pigment. The connection between retinulae and nerves is well seen.

Fig. 43. Cornea, cone, retinula and optic ganglion from the lateral eye of *Phronima.* × 157.

Fig. 44. *A* cone from the dorsal eye of *Phronima.* The thread-like extension is not shown in its full length. × 157.

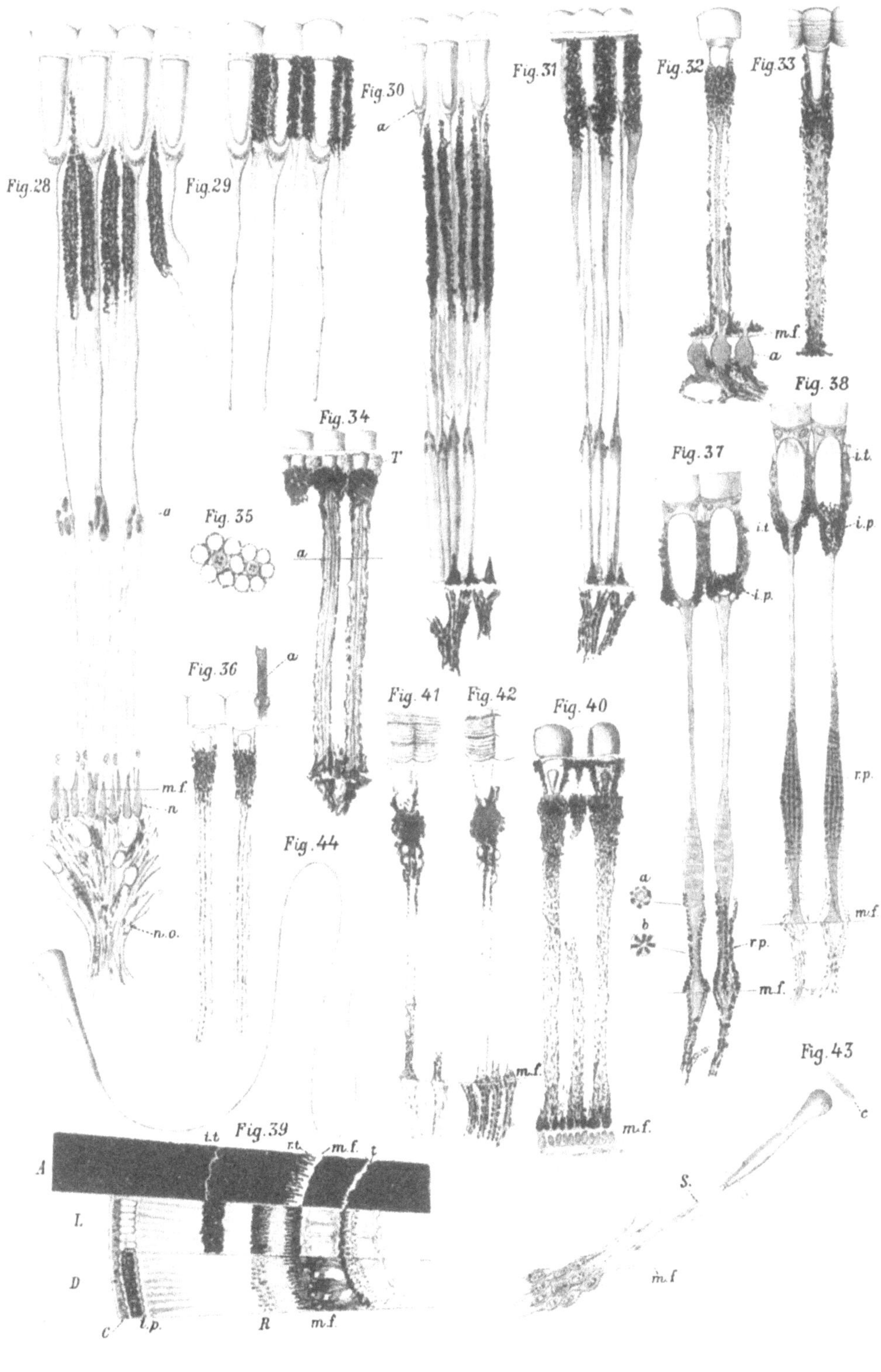
Fig. 28
Fig. 29
Fig. 30
Fig. 31
Fig. 32
Fig. 33
Fig. 34
Fig. 35
Fig. 36
Fig. 37
Fig. 38
Fig. 39
Fig. 40
Fig. 41
Fig. 42
Fig. 43
Fig. 44
a
m.f.
n
n.o.
T
i.t.
i.p.
r.p.
b
c
S.
A
I.
D
C
R
r.t.
t

Plate V

In this plate tapetal substance is coloured yellow. This is not the case in reality, but serves to indicate that this substance appears dark, like pigment, in transmitted light, but, in contrast to pigment, bright in incident light.

Figs. 45, 46. Ommatidia from *Galathea.* ×92. **Fig. 45** dark- adapted, **Fig. 46** light-adapted. Both retinal and iris pigment have migrated. Fixed in Müller's solution.

Fig. 47. Ommatidia from *Peneus membranaceus.* ×114. No pigment seen at all, but only the usual three tapetal layers. Fixed in alcohol. (As described in the text, I did find traces of pigment in another individual of this species) *i.t.* iris tapetum *r.t.* retinal tapetum *t.* tapetal layer in optic ganglion *m.f.* Membrana fenestrata.

Figs. 48, 49. Dark- and light-adapted eyes of *Pisa.* ×166. *m.f.* Membrana fenestrata *a, b, c, d, e, f* cross-sections at higher magnification at the levels indicated. The drawing is composed partly from animals fixed in alcohol and partly from those fixed in Müller's.

Fig. 50. Ommatidia of a spiny lobster (*Palinurus*). ×46. The part of the retina to the right of the line shows the retinal pigment in the dark-adapted state. Fixed in Rabl's solution. *a, b, c, d* cross-section through the levels indicated. *c & d* at higher magnification.

Figs. 51, 52. Dark- and light-adapted eyes of *Palaemon.* ×146. Exquisite migration of both pigment layers. Fixed in Rabl's and post treated with alcohol. *c.* cornea *i.t.* iris tapetum *i.p.* iris pigment *r.p.* retinal pigment *r.t.* retinal tapetum *m.f.* Membrana fenestrata *t.* tapetal layer in optic ganglion.

Figs. 53, 54. Dark- and light-adapted eyes of *Sicyonia sculpta.* ×124. Fixed in alcohol.

Figs. 55, 56. Dark- and light- adapted ommatidium of *Dromia vulgaris.* ×104. The animal of **Fig. 56** was only kept in diffuse daylight. Fixed in alcohol, post-treated (decalcification) with Rabl's.

Fig. 57. Crystalline cone and corneal facet of the moth *Leucoma salicis* as seen in the microrefractometer to show the dioptric properties. The instrument's screen is on the right. 1. Corneal facet in which a core stands out from the marginal zone by virtue of its higher converging power (as a result of layering) 2. A drum-shaped section of a cone (cut distally and proximally) still shows converging power (lens cylinder). Surrounded by its sheath 3. Cone and sheath seen from the side. In the region of its tip it is surrounded by pigment and therefore appears dark on both sides.

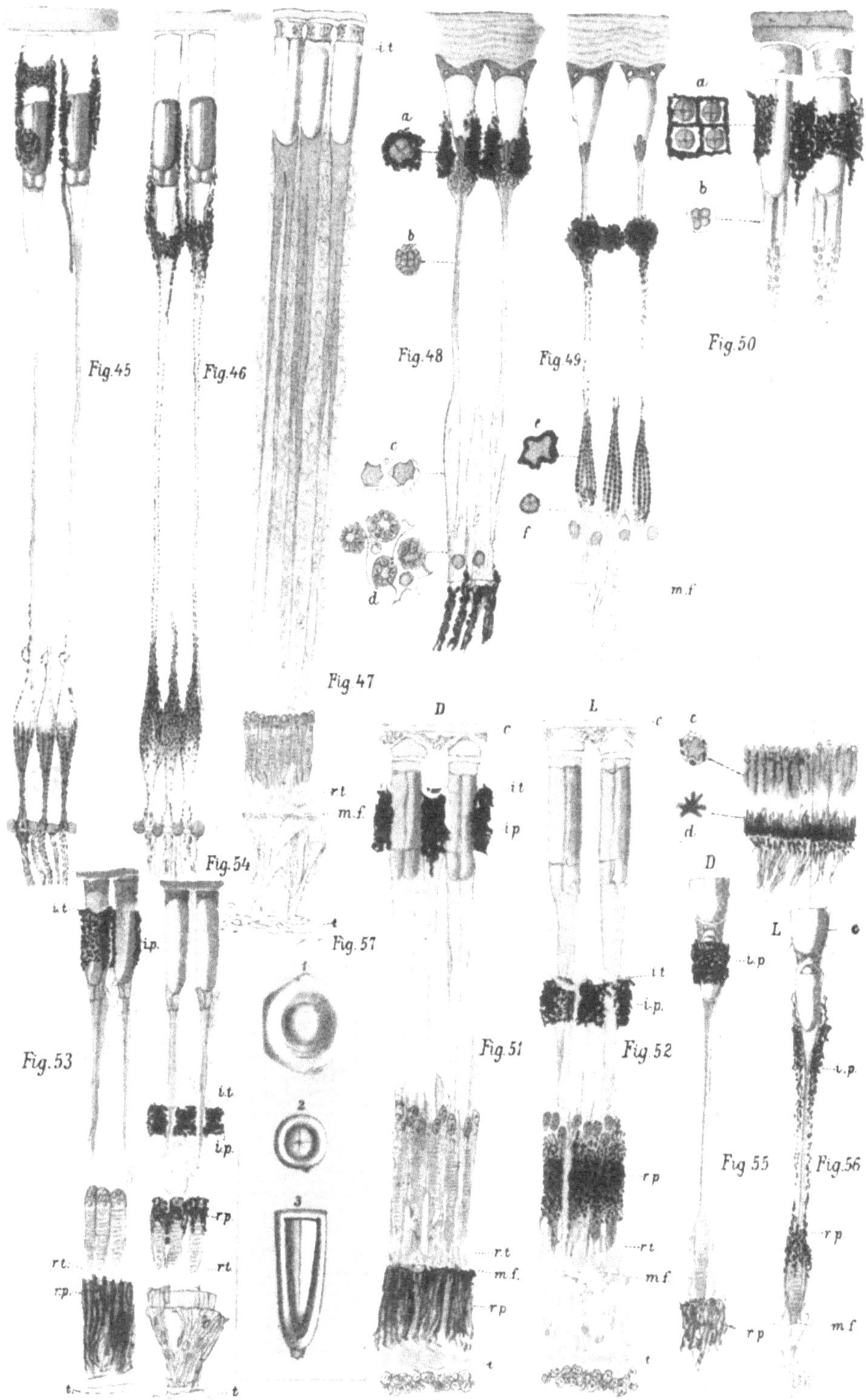
Fig. 45
Fig. 46
Fig. 47
Fig. 48
Fig. 49
Fig. 50
Fig. 51
Fig. 52
Fig. 53
Fig. 54
Fig. 55
Fig. 56
Fig. 57

Plate VI

Figs. 58, 59. Each shows two ommatidia, the first from the dorsal and the latter from the ventral eye half of *Libellula vulgaris* (Diplex). × 158. Notice the difference in size and the lack of black pigment in the dorsal eye. *m.f.* Membrana fenestrata *t.* cross section through trachea.

Fig. 60. Ommatidia of a fly: *Eristalis* × 180
a. A hair *b.* Tracheae *c.* cross-section of the retinulae (which appear dark) with thick tracheae lying between.

Fig. 61. The eye of the cabbage white (*Pieris rapae*) in vivo viewed with a lens. (Figs. 34, 35 show sections of the same eye). The largest black spot is the principal pseudopupil in the middle of which a red glowing spot would be seen with direct sunlight and an ophthalmoscope. Six first order accessory pseudopupils surround the principal pseudopupil – somewhat irregularly. Still further are second order accessory pseudopupils with no obvious regularity.

Fig. 62. Eye of the painted lady (*Vanessa cardui*) in vivo seen with the lens. Principal pseudopupil with six first order accessory pseudopupils and a few second order ones still noticeable near the eye margin.

Fig. 63. Eye of the humming-bird hawkmoth (*Makroglossa*) in vivo seen with a lens. The principal pseudopupil is visible. The first order accessory pseudopupils are found at the eye margin. Although six in number they cannot be confidently seen from one direction of observation. With ophthalmoscope and sunlight the centre of the principal pseudopupil would appear to glow green.

Fig. 64. *Epinephele*, in vivo, observed with a lens. A dark bar connects the principal pseudopupil to two of the first order accessory pseudopupils, which in turn are connected to each other by a dark ring. There are indications of second order accessory pseudopupils.

Fig. 65. Eye of *Galathea*, in vivo observed with a lens in diffuse light. The principal pseudopupil is square, corresponding to the square facets. Glimmering lines criss-cross the eye, parallel to the sides of the pupil. Their regularity and fineness are not adequately reproduced.

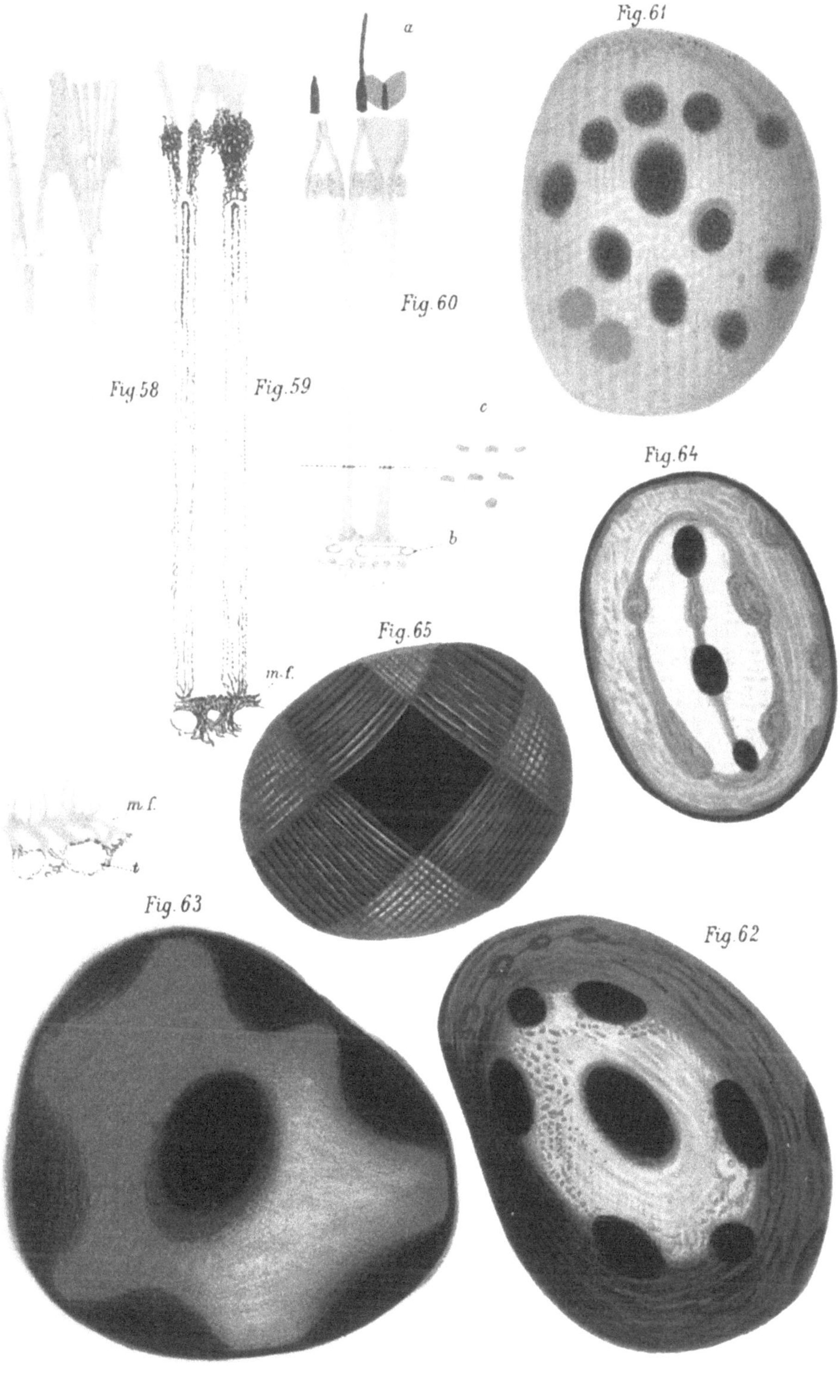
Fig. 61
a
Fig. 60
Fig 58
Fig. 59
c
Fig. 64
b
Fig. 65
m.f.
m.f.
t
Fig. 63
Fig. 62

Plate VII

Fig. 66. Eye of a large dragonfly (*Cordulegaster*) viewed from medial ventral direction with the naked eye. Principal pseudopupil, partially fused first order and a number of higher order accessory pseudopupils.

Fig. 67. Eye of the same dragonfly seen from above. Shows the large principal pseudopupil of the dorsal eye and a number of first and second order accessory pseudopupils in the the lateral part of the eye.

Fig. 68. Eye of a dark-adapted *Carcinus maenas* seen with the ophthalmoscope and lens. The elongated pseudopupil has a red glowing centre (luminous pseudopupil) and is surrounded by a bright halo. The spots surrounding the phenomenon are unrelated and do not move with the direction of observation. They are pigment spots.

Figs. 69, 70. Light- and dark-adapted eyes of *Palaemon* seen with the ophthalmoscope. Following a few minutes of illumination the appearance of **Fig. 70** transforms to that of **Fig. 69**. Sections through these eyes are shown in **Figs. 51, 52**.

Figs. 71, 72. Light- and dark-adapted eyes of *Pagurus* viewed as above. **Fig. 71** shows a few just noticeable first order accessory pseudopupils some of which appear doubled (second order accessory pseudopupils?)

Figs. 73, 74. Light- and dark-adapted eyes of *Sicyonia sculpta* viewed as above. Sections of these eyes shown in Figs. 39, 53, 54.

Fig. 75. The eye of *Peneus membranaceus* in diffuse daylight viewed with lens only. The pseudopupil forms the centre of a four-armed star. Section of this eye in Fig. 47

Fig. 76. The eye of *Nica edulis* with lens only but seen under water. With ophthalmoscope and sunlight the black pseudopupil – or its centre glows. The white stripes represent a glimmering reflection the fine detail of which is inadequately reproduced.

Fig. 66

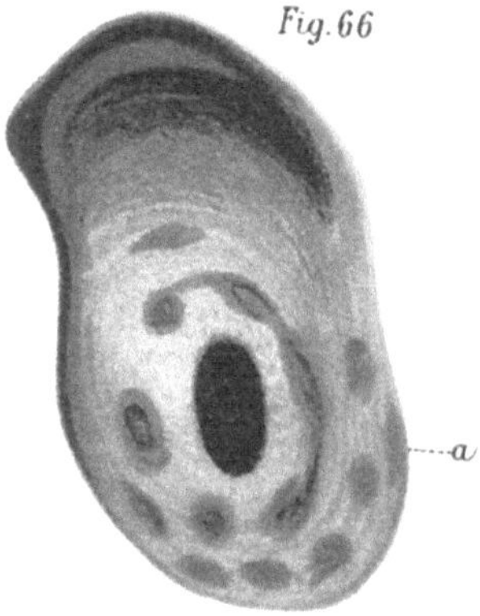

Fig. 67

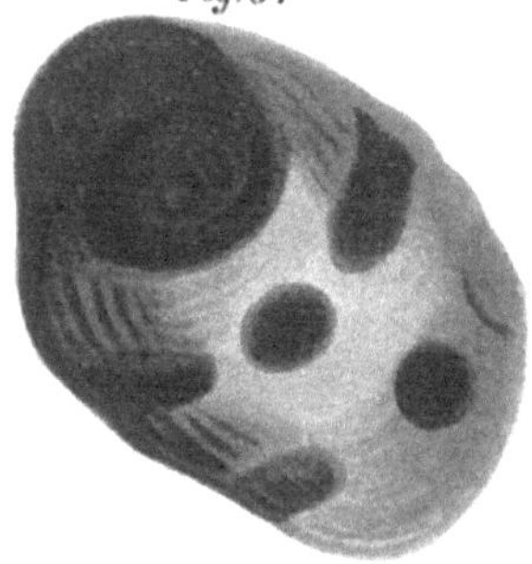

Fig. 68

Fig. 69

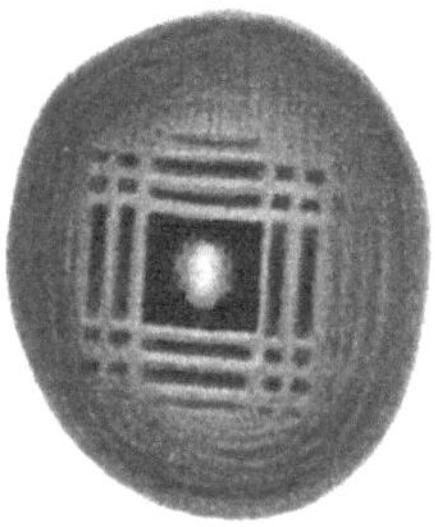

Fig. 70

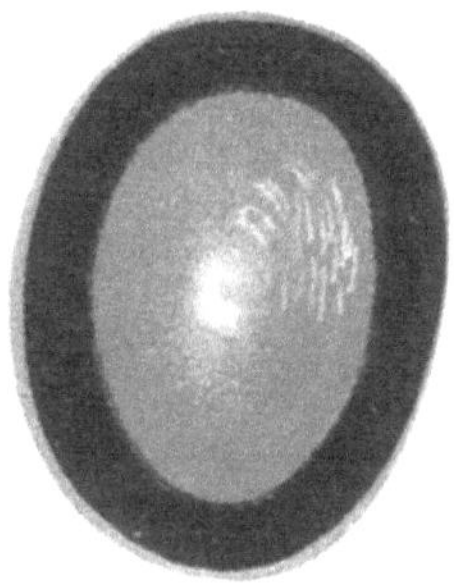

Fig. 75

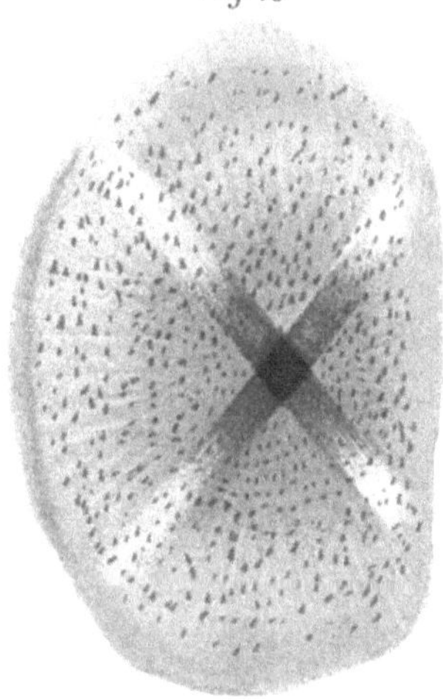

Fig. 71

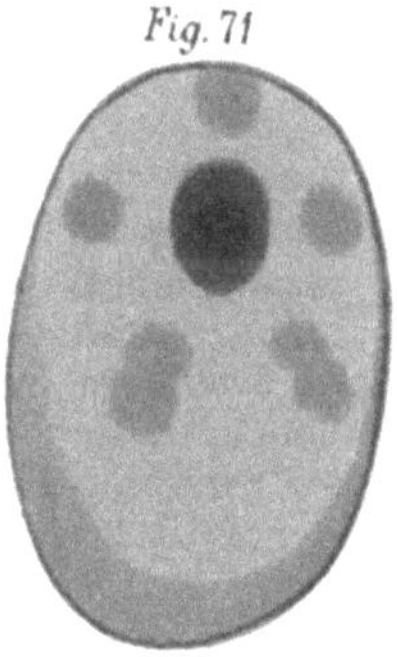

Fig. 72

Fig. 73

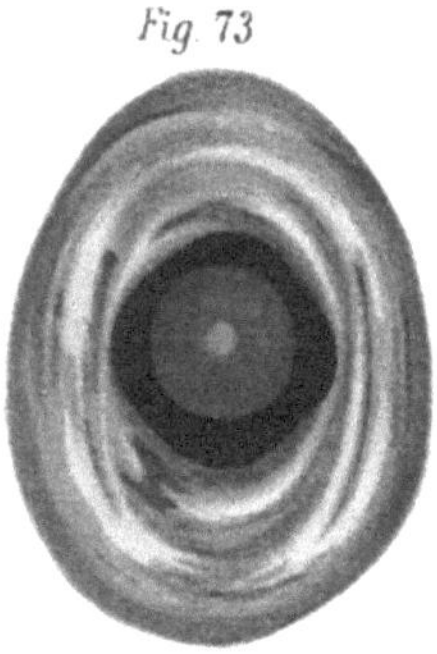

Fig. 74

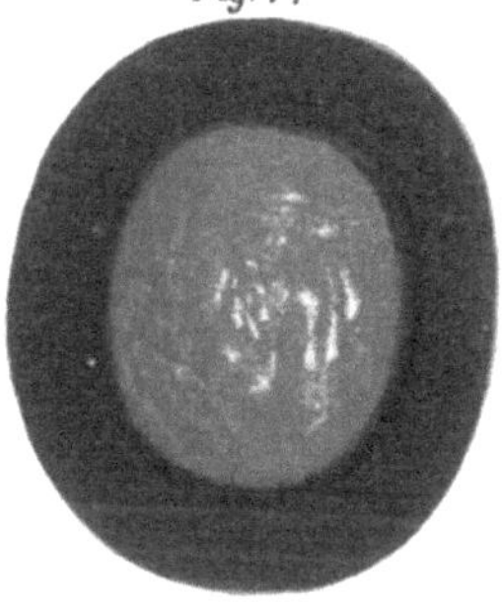

Fig. 76

Notes

Chapter I

1. *(p. 2)* Fletcher et al. (1954) derived a more precise expression for the ideal refractive index gradient for an imaging lens cylinder:

$$n_r = n_o \operatorname{sech}(\pi r/2F),$$

where n_r is the refractive index at a radial distance r from the axis, n_o is the axial refractive index and $2F$ the distance between foci.

As discussed by Land (1980) the parabolic function of Exner's brother Karl is in fact a very close approximation to this. Actual measurements of refractive index gradients in lens and crystalline cones of superposition eyes are also quite adequately fitted by Exner's parabolic approximation (Seitz 1969; Hausen 1973; Vogt 1974; rev. Kunze 1979).

The discovery of the lens cylinder principle rates as one of Exner's major contributions, and is probably the first example of compound eye research which has led to a technical application. Although Schott manufactured glass lens cylinders at Exner's instigations (Exner 1886), only in very recent years have lens cylinders been produced commercially. Their property of periodically focusing (as a function of length) can be exploited for the transmission of images as an alternative to coherent fibre optics.

2. *(p. 4)* Whether this potential advantage of the lens cylinder is ever exploited seems debatable. The divergence of the axial rays (after the focus) is generally speaking much greater than the extra degree or so introduced by consideration of paraxial rays, and this divergence is as great – or greater – for a lens cylinder. In both cases (lens/lens cylinder) it should be appreciated that, at least for apposition eyes, there is little point in having an f number (which determines the degree of divergence) so small that the divergence exceeds the critical angle of internal reflection of the rhabdom light guides. In the fly (*Musca*), where the image is produced by a spherical surface (outer corneal surface), rays reaching the rhabdomeres from the very edge of the lens just exceed the critical angle for internal reflection. This can be seen with the light path reversed (antidromic illumination) since the facet lens is not quite completely illuminated (Stavenga 1975). For a more rigorous discussion using wave guide optics, see van Hateren (1985).

Chapter II

3. *(p. 8)* Exner is referring to the fact that Gottsche used a fly cornea where the cones were missing so that he could not be certain that such an image really occurred in vivo (discussed in more detail by Grenacher 1879). Even so we now know that an inverted image is formed by facet lenses in the fly – and that this is almost entirely due to the corneal lens outer surface (as Exner himself p. 60 demonstrates). In the fly the inverted image is also actually analysed by the seven rhabdomeres – the one case where the theories of Gottsche, Schultze etc. are, in a sense, valid. However, the final effect is still of one "perception" per ommatidium (or neuro-ommatidium) thanks to the neural superposition principle (Vigier 1907a, b, c; Kirschfeld 1967; Braitenberg 1967 revs. Braitenberg and Strausfeld 1973; Hardie 1985; see Appendix I).

4. *(p. 10)* Retinula: Apart from the word "Retinula", Exner also uses the word "Sehstab" (lit. visual rod) which is almost always synonymous with his use of Retinula. Both have been translated as retinula.
5. *(p. 11)* Apparently the limits on acuity posed by diffraction were not considered by Exner. We now appreciate that the diameter (D) of the lens (facet) is the principle limiting factor for photoreceptor acuity. The angular size of the Airy disc (blur circle of a point source in the focal plane of a lens),

$$A = \lambda / D,$$

where λ = wavelength of light. This sets a limit to the acuity of the underlying photoreceptors. (see e.g. Barlow 1952; de Vries 1956; Kuiper 1966; Snyder 1977; Land 1981a). Mallock (1894) seems to be the first to have realized the importance of this relationship in compound eyes. He pointed out that if Müller's theory of mosaic vision was correct, then the diffraction limit of the facet lens should be more or less matched to the interommatidial angle. Using anatomical measurements of facet diameter and interommatidial angles he found a close agreement in a number of species and thus a) confirmed Müller's theory and b) gave quantitative estimates of resolution and acuity which were $\pm$ 50% accurate.
6. *(p. 15)* This rough estimate of the acceptance angle is approximately correct. Recent measures of acceptance angle from intracellular recordings of photoreceptors in the *Limulus* lateral eye (rev. Barlow et al. 1989) reveal values of 6° (light-adapted) and 13° (dark-adapted).
7. *(p. 17)* As Exner correctly points, out this is an elegant and widespread design feature of compound eyes. It is also possible to skew the ommatidial axes in the opposite sense and to create higher resolution (greater density of ommatidial axes). This was realized by Demoll (1917) and developed further by Stavenga (1979), both of whom discuss it as a design strategy for creating a fovea (zone of acute vision).
8. *(p. 19)* In fact we now know that internal reflection is widely used in those eyes with reflecting superposition optics (Vogt 1975, 1977, 1980; Land 1976; revs. Land 1981a; Nilsson 1989; see Appendix I)
9. *(p. 20)* These are the cone cells.
10. *(p. 20)* see note **2**.
11. *(p. 20)* H. Müller (1854) determined the location of the sensitive layer in the human retina most elegantly, by exploiting the image of blood vessels seen when light is shone into the corner of the eye. By illuminating from different angles and measuring the parallax of the image, he calculated that the sensitive layer must correspond to the outer segments of the rods and cones.
12. *(p. 20)* This seems to be a reference to the concept of an image sweeping across the rhabdom. This concept (interaction between rhabdom and dioptrics) has since blossomed into the field of photoreceptor optics, which allows a quantitative appreciation of angular sensitivity etc. in photoreceptors. Apart from the odd statement like this, Exner seems virtually to have ignored this important factor (revs. see Snyder and Menzel 1975; Snyder 1979; van Hateren 1989).
13. *(p. 21)* Brücke (1844), supposed that the rods/cones acted as light guides, but believed that the sensitive layer of the retina was in front of (distal to) the outer segments. Following the view of Bidder (1839), he argued that, in animals with a tapetum, the outer segments allowed reflected light to be guided back directionally so that the resolution of the image was not impaired.
14. *(p. 22)* Significant changes do indeed occur in the *Limulus* eye: at night the pigment aperture at the cone apex opens up, and the retinula cells move closer to the cone apex, the cone cells becoming compressed in the process. The changes are associated with a large increase in sensitivity (100 ×) and a concomitant decrease in receptor acuity (2 ×). The changes are under the control of a circadian rhythm mediated by efferent nerve fibres (rev. Barlow et al. 1989).
15. *(p. 22)* *Limulus*, the only living representative of the Xiphosura (phylum Chelicerata), is considered the nearest living relative of the trilobites.
16. *(p. 23)* Theories of trilobite vision inevitably remain speculative. Most trilobites had "holochroal" eyes which were similar to modern-day compound eyes in having a closely packed lattice of lenses and ommatidia. *Phakops*, which was studied by Exner, belongs to the suborder Phacopina, which had schizochroal eyes with very large lenses (sometimes more than 250 μm) separated by sclera. Modern opinion supports Exner in suggesting that a discrete retina may have subserved each lens. The lenses of these eyes apparently had a doublet construction

resulting in correction for spherical aberration (Clarkson and Levi-Setti 1975). The visual fields of each lens overlap strongly with neighbours above and below, but not horizontally. This has led to the suggestion that monocular depth perception may have been performed between vertically aligned neighbours, using vertical disparities (Stockton and Cowen 1976; rev. Schwind 1989).

17. *(p. 23)* The terms superposition eyes/apposition eyes are used throughout the translation although Exner himself never used them, always referring – as here – to superposition/apposition images. The terms were already in widespread use a few years after Exner's monograph (e.g. Demoll 1917).

18. *(p. 24)* Vogt (1974) pointed out that glycerine solutions tend to permeate the dioptrics of the eye thus significantly reducing refractive index gradients. This behaviour probably accounts for some of the quantitative discrepancies in Exner's results. Solutions of bovine albumin are preferred by Vogt as the molecules are too large to invade the optical media.

19. *(p. 26)* (see note **18** above).

20. *(p. 42)* In fact the metallic shine of many insect eyes (particularly tabanid flies) is due to interference reflection colours arising from alternating dense and rare layers located just under the corneal surface. The layers act as quarter wavelength reflectors, the colour of reflection thus being determined by the exact separation of the layers. (Bernard and Miller 1968; rev. Miller 1979).

Chapter III

21. *(p. 43)* Due to the great diversity, the classification of pigment cells is rather confused. In insects, the cells containing Exner's "iris pigment" include both primary pigment cells, which generally surround the crystalline cone and the distal retinula, and also accessory or secondary pigment cells which usually traverse the entire retina. Exner's "retinal pigment" resides either in the proximal portions of the same secondary pigment cells or in extra basal pigment cells. In crustaceans the terms distal and proximal pigment cells are usually preferred. Exceptions abound, however (see also, Trujillo-Cenoz 1972; Autrum 1981; Hardie 1985; Stavenga 1989).

22. *(p. 43)* This region has been called the "clear zone" (Horridge 1971). During the 1960–1970's when the existence of superposition images in any eyes became questioned, there was a move to reclassify superposition eyes as "clear-zone" eyes, but most authors now use the term "superposition eyes".

23. *(p. 44)* The maximum number of facets contributing to a superposition image (and hence the maximum aperture change) is ca. 500, (Höglund 1965) but often less (e.g. 100–200 in *Ephestia*, Kunze 1979). Whilst diurnal vertebrates such as man can change the area of their aperture by only a factor of ca. 16, in nocturnal vertebrates much larger changes are possible when the closed pupil takes on the form of a slit. The gecko is probably one of the most extreme examples, but even the cat can change its aperture by a factor of 135. (see e.g. Walls 1942; rev. Hughes 1977).

24. *(p. 44)* The iris pigment has since been shown to migrate in a number of apposition eyes: e.g. praying mantis (Rossel 1983), many crabs and *Limulus* (see note **14** above), with similar effects on acuity and sensitivity (rev. Stavenga 1979). In addition the majority of apposition eyes display migration of pigment granules in the photoreceptors themselves – the granules taking up a position close to the rhabdom in the light where they reduce the light flux (Kirschfeld and Franceschini 1969; rev. Stavenga 1979).

25. *(p. 45)* We now know many examples where pigment migration is also under control of a circadian rhythm (e.g. *Limulus*, crabs, mantis rev. Barlow et al. 1988). Exner apparently did not consider this – or control for it in his experiments.

26. *(p. 46)* As in other superposition eyes, the clear zone in *Hydrophilus* is traversed by extensions of both pigment cells and retinula cells, and movements on light and dark adaptation presumably reflect movements of these cells. No muscle fibres are present here or in any other compound eye. Exner's confusion may be explained by Horridge and Giddings' (1971a) observation that, in aquatic beetles, the retinula cells can hardly be distinguished from the pigment cells. The distal cone-shaped processes Exner described may have represented the distal rhabdomeres, which are particularly large in *Hydrophilus* (Horridge and Giddings 1971a).

Chapter IV

27. *(p. 54)* Superposition eyes are also found in some day-flying moths (the hummingbird hawkmoth *Macroglossum* is in fact discussed by Exner). A recent study (Land 1984) shows that such eyes (e.g. in the agaristid, *Phalanoides*) can approach the acuity of apposition eyes, but with greatly superior sensitivity.

28. *(p. 56ff)* Exner spends five pages exhaustively considering whether the macrurans have a) lens cylinders and b) superposition images. It is almost as if he suspected that something was not quite right! In fact it is precisely these crustaceans which employ the newly discovered principle of reflecting superposition whereby image-forming rays are reflected of the walls of the cones (Vogt 1975, 1980; revs. Land 1980; Nilsson 1988a; Appendix I & II). The crystalline cones in fact are not lens cylinders, and Exner was probably confused by dense margins which contain crystals which function as multi-layer reflectors. The lack of a refractive index gradient in these crystalline cones (e.g. Kuiper 1962) along with the discovery of crystalline threads (see note **32** below) almost led to the abandonment of the superposition theory in the 1960's (revs. Horridge 1975; Land 1981a; Nilsson 1989a).

29. *(p. 57)* Many of the transitional forms described by Exner in fact belong to a new class of superposition eye – the parabolic superposition eye (see Appendix 1; Nilsson 1988). In these eyes, which are characterized by the presence of light guide structures crossing the clear zone, the superposition image is formed by reflection off the walls of the crystalline cone. However, unlike the reflecting superposition eye, the walls have a parabolic curvature which optically corrects for the convergence of rays caused by the corneal lens. Numerous examples of analogous transitional forms (between apposition and classical refracting superposition eyes) are also found in insects, particularly amongst beetles (click beetles and soldier beetles – Caveney and McIntyre 1981).

30. *(p. 60)* 30 cm in original: presumably this was a printing error.

31. *(p. 60)* Measurements show that the pseudocone in the fly (*Calliphora*) in fact has a slightly lower refractive index than that of the surrounding (primary) pigment cells (Seitz 1968). As far as the focal length is concerned the effect is negligible. Seitz argues that this does have the consequence that non-axial rays striking the sides of the cone are more likely to be refracted into the pigment cells (and thus absorbed) than to be reflected – thus impairing the image.

32. *(p. 61)* These vestiges, composed variously of rhabdomere threads and crystalline tracts (formed by the secondary pigment cells), fed fuel to the controversy in the 1960's and 1970's over whether a superposition image was ever realized. Thus it was supposed that they might act as light guides, guiding light to the spindle-shaped rhabdoms – the eyes thus acting as apposition eyes (e.g. Horridge 1969). It is now generally accepted that Exner was correct. They may perform this function for axial rays in the light-adapted state, but in the dark-adapted eye a superposition image is indeed formed (revs. Horridge 1975; Land 1981a; Nilsson 1988; see also note **29** above).

Chapter V

33. *(p. 64)* This is still the most widely accepted definition of the arthropod retina, although some authors do use a more vertebrate-oriented terminology, referring to the receptor layer as the outer plexiform, and the lamina as the inner plexiform layer (e.g. Trujillo-Cenoz 1972).

34. *(p. 65)* It is interesting that, although Exner considered (like Grenacher) that the rhabdom was probably the site of transduction (see e.g. p. 64), he was prepared to accept that it was a cuticular formation. This view was still put forward by some as late as the 1950's (one theory e.g. being that the rhabdom was simply a structure designed to distribute incident light into the cytoplasm of the retinula cells where transduction took place – Nowikoff 1931). This idea was only finally laid to rest with the first EM pictures of the rhabdom (Fernandez-Moran 1956; Danneel and Zeutschel 1957; Miller 1957; Goldsmith and Philpott 1957) showing that the rhabdomeres consisted of microvilli, and the first microspectrophotometry (Langer 1965) showing that they contained a rhodopsin-like pigment. In comparison it was already clear in the last century that the rods of the vertebrate retina contained visual pigment (H. Müller 1854; Boll 1876; Kühne 1878).

35. *(p. 66)* The butterfly tapetum is an exception: here light is reflected by a "private" tracheal tapetum underlying each rhabdom. The colour of the reflection indicates that the tapetum is acting as an interference reflector. (Miller and Bernard 1968; rev. Miller 1979)

36. *(p. 66)* The guanine crystals of certain tapeta in vertebrates and also *Limulus* have a refractive index of 1.83 which in fact generates an even better reflectivity than with an air interface (rev. Miller 1979).

37. *(p. 67)* see note **36** above

38. *(p. 69)* See note **21** for discussion of pigment cell nomenclature.

39. *(p. 70)* Although movements of the distal pigments are more obvious, examples of proximal retinal pigment movement are also found in insects (e.g. Horridge et al. 1983). In addition, there are extensive retinomotor movements in both crustaceans and insects, whereby the retinula cells themselves migrate, often along with the pigment cells (rev. Walcott 1975). Two reasons may have contributed to Exner's failure to make these observations: firstly not realizing that a circadian rhythm is involved in pigment migration, and secondly, extreme light and dark-adapted states are often only reliably obtained when eyes are fixed in boiling water (Horridge and Giddings 1971a).

40. *(p. 71)* Exner correct: the retinula cells themselves contain pigment, and there is also tapetal pigment.

41. *(p. 73)* Extensive tapetal migration has been described for example in the grass shrimp, *Palaemonetes* as well as changes in reflecting platelets in the distal pigment cells (i.e. Exner's iris pigment cells) (Doughtie and Rao 1984).

Chapter VI

42. *(p. 76)* Varying degrees of regional specializations are the rule rather than the exception. Particularly striking examples are found in the males of many species (e.g. mayflies and many dipterans) which use especially enlarged dorsal eye regions for detecting and tracking females (Pictet 1843; Dietrich 1909; revs. Horridge 1977; Wehner 1981; Land 1989).

43. *(p. 77)* See also note **5**: the finely facetted pattern is in fact almost always a mark of poor acuity!

44. *(p. 78)* Exner's arguments on form and motion detection were flawed: see note **71**.

Chapter VII

45. *(p. 79)* These rods are not in fact the rhabdomeres, but concentrations of red pigment granules in the retinula cells which abut four of the rhabdomeres, thus acting as a spectral filter. The rhabdom itself is a typical fused rhabdom (Ribi 1978).

46. *(p. 79)*. These oval structures beneath the basement membrane probably represent the giant nuclei of the basal pigment cells *(Yagi and Koyama 1963)*.

47. *(p. 83)* All the macrurans (*Palaemon*, etc.) investigated by Exner have reflecting superposition eyes.

Note on taxonomy: Crustacean taxonomy has been extensively revised since Exner's day: those crustaceans which Exner refers to as "Langschwänze" (translated in text as macrurans) are all long-bodied crustaceans which are now split between two suborders (Dendrobranchiata and Pleomyceta). His "Kurzschwänze" correspond to brachyurans – the true crabs (an infraorder of the Pleomyceta), and the "Halbschwänze" correspond to the anomurans (squat lobsters and hermit crabs), another infraorder of the Pleomyceta.

48. *(p. 84)* The striations Exner described (first seen by Schultze) come from the interdigitation of the constituent rhabdomeres, each layer having a microvillar orientation orthogonal to the next. This organization maintains the effective dichroism of the rhabdomere along its length. Nevertheless, the normal fused rhabdom found in most insects can perform this function just as well. To date there is no satisfactory theory suggesting any additional advantage of the interdigitated organization in crustaceans (rev. Waterman 1981; Doujak 1984).

Chapter VIII

49. *(p. 88)* As usual, Exner recognized the salient features of these unusual eyes. It is worth mentioning that *Squilla* is a "mantid shrimp", a predatory species which catches its prey, like the praying mantis, by rapid extension of its forelimbs. There is thus a definite behavioural requirement for a good range finding mechanism. Since the eyes are on stalks, accurate judgement of distance by binocular cues may be subject to error (see also Horridge 1977). A recent study (Marshall 1988) reveals a remarkable further subdivision of function in the 6 ommatidial rows comprising the central band in two related stomatopods, *Odontodactylus* and *Gonodactylus*. Rows 2 & 3 each have a variety of coloured spectral filters in the rhabdomere suggesting a colour vision system in which the different spectral classes of photoreceptors are generated by a filter mechanism rather than different visual pigments. The microvillar arrangements in rows 5 & 6 suggest a specialization for the detection of polarized light. Ommatidial rows outside the central band, but with aligned optical axes, can be expected to be involved in a monocular range finding mechanism.

50. *(p. 89)* Land (1981b, rev. 1989) has made a detailed study of the medial eye in *Phronima*, calculating interommatidial angles from the pseudopupil and estimating acceptance angles of single cones by ray tracing after determining the refractive index gradient in the cones by interference microscopy. He found that interommatidial angles were exceptionally small (0.44°), whilst acceptance angles are 8 × larger ($\Delta\varrho = 3.5°$). The large overlap of receptive fields is interpreted as an adaptation for detecting small low contrast objects (e.g. prey) against a homogeneous background under dim light conditions. A necessary prediction of this design strategy is that overlapping receptors are pooled. *Phronima* is a deep sea amphipod which lives in a "barrel" formed from the case of certain tunicates. Both the barrel (with which *Phronima* swims around searching prey or mates) and the animal itself are basically transparent apart from the retinae. The ploy of concentrating the retina as a small spot at the base of long tapering optical elements can thus be seen as an adaptation for minimizing visibility (see also Nilsson 1989).

51. *(p. 93) Copilia* is another deep sea crustacean (collected at 200 m by Gregory et al. 1964) which is also largely transparent (thus again a tiny retina can be seen as advantageous). According to Wolken and Florida (1969), the focal length of the anterior lens is considerably longer than the distance to the posterior cone. The cone would thus further focus an only partially focused image onto the rhabdomeres (5 in number). The result is an effective decrease in the f number (from 5.4 to 1.6), i.e. a considerable increase in the intensity of the image (rev. Land 1981a). A clue to the possible function of this extraordinary eye is given by the mating behaviour. Female *Copilia* locate the males by their flashing phosphorescent signal (Gregory 1986). Since there is a marked sexual dimorphism (whereby the female eyes are much better developed), it is possible that this is the main use to which the eyes are put. Perhaps a visual world consisting of a point source is adequately served by a point retina.

Chapter IX

52. *(p. 99)* Schultze (1868) originally described the laminated structure of crustacean rhabdoms, and considered this analogous to the laminations he had seen in rods and cones (in fact due to the tendency of the discs to break off in clumps). In this he saw a fundamental structure – essential to photoreception – which gave rise to interference reflections. Zenker (1867) suggested a theory for colour sensitivity based on this – namely that specific interference colours where reflected as a function of the separation of the laminae.

53. *(p. 101)* For an apposition eye this can be true (when the size of the glow is measured appropriately as the far-field radiation pattern (Franceschini 1975). However, for a superposition image the size of the eyeshine is determined principally by the number of facets contributing to the superposition image. This can only slightly be influenced by any scattering in the retina.
54. *(p. 103)* There appears to be no explanation in the literature for these colour changes, though they are presumably due to filtering effects of one or other of the pigment layers. Any photochemical effects on the visual pigment would be too minor to be seen.
55. *(p. 104)* This phenomenon has been observed repeatedly (e.g. Demoll 1917). The full explanation is quite recent: at the base of each rhabdom there is a trachea which functions as an interference reflector. Different colours are reflected as a function of the separation of the layers in the trachea. On light adaptation pigment granules in the retinula cells migrate to the rhabdom thus attenuating the light flux and observed glow (Miller and Bernard 1968; rev. Miller 1979).
56. *(p. 105)* The size is a measure of the aperture of the viewing system, and the interommatidial angle. Only in the far-field radiation pattern (observed with telescopic optics) does the size of the glow from a single facet give a measure of the photoreceptor's acceptance angle in apposition eyes (Franceschini 1975; van Hateren 1984).
57. *(p. 105)* The retinulae appear as red threads because of dense red screening pigment in the retinula cells (Ribi 1978). It is thus no wonder that they did not bleach. (See also note **45**)
58. *(p. 106)* Examples of beetles with eyeglow have since been described (e.g. *Repsimus* – Horridge 1971).
59. *(p. 106)* Despite numerous attempts it has not proved possible to reproduce this observation of eyeshine in dragonflies (DG Stavenga personal communication).
60. *(p. 107)* The faster movement over the dorsal surface simply shows that the interommatidial angles are smaller (thus indicating better resolution). When such observations are quantified, maps of the resolution over the eye can be very simply generated (e.g. Horridge 1977; Stavenga 1979). However, this allows no conclusion about motion detection!
61. *(p. 111)* (See also notes **14, 39**). The mechanisms of pigment migration in compound eyes are still only poorly understood (revs. Autrum 1981; Stavenga 1989). It is generally assumed that the mechanism for both pigment migration and gross cellular movements involves the cytoskeleton, and in particular the often extensive arrays of microtubules. It has also been suggested that the migration of rhabdomeres may be accomplished by a process of renewal at one end and breakdown at the other (for review and references see Walcott 1975; Autrum 1981). In insects direct actions of light on pigment migration in both retinula cells (Kirschfeld and Franceschini 1969) and pigment cells (Hamdorf and Höglund 1981; Land 1986) have been demonstrated. Evidence exists for calcium as the necessary intracellular messenger in retinula cells (e.g. Kirschfeld and Vogt 1980). In crustaceans pigment-concentrating and -dispersing hormones have been identified and parallels with the control of pigment dispersal in chromatophores have been made (revs. Rao 1985; Stavenga 1989).
62. *(p. 112)* Even fourth order accessory pupils may be seen in some hymenopterans (Stavenga 1979).
63. *(p. 113)* Exner is mistaken here: all accessory pseudopupils move in synchrony. Exner is probably referring to what Stavenga (1979) calls the oblique illumination pseudopupil, a dark spot which can arise in ommatidia aligned with light source but not the observer.
64. *(p. 122)* Exner's interpretation of the accessory pseudopupils is essentially correct – they represent the proximal (iris) pigment observed through the facet lenses of neighbouring ommatidia. The pattern the image reveals the place of the pigment cells in the lattice structure of the compound eye (Franceschini and Kirschfeld 1971; Stavenga 1979).

Chapter X

65. *(p. 123)* Snellen ratio: this measure of acuity is the distance at which a test object is resolved divided by the distance at which the same object would subtend 1 minute of an arc. 6/500 thus corresponds to ca. 1.7° (probably a little optimistic for *Lampyris*).

66. *(p. 123)* The sharpest superposition eye yet described comes from a day-flying moth (the agaristid *Phalanoides*). With a receptor acuity of ca. 1–2° it approaches the performance of comparable apposition eyes (Land 1984).
67. *(p. 124)* In terms of resolution as determined by receptor or ganglion cell spacing, large distortions also occur in vertebrate retinae (viz. visual streaks, rev. Hughes 1977).
68. *(p. 125)* Since it is the absolute length and optical density of the waveguide structure (rhabdomere, rod or cone) which determines the proportion of incident light absorbed, small animals cannot afford to have shorter waveguides. In fact the microvillar organization of rhabdomeres seems to be slightly less efficient at absorbing incident light than the stacks of discs in vertebrate rods. This may explain why many arthropod retinae are even thicker than vertebrates (Kirschfeld 1969).
69. *(p. 126)* There is still no really satisfactory explanation for *why* some animals have compound eyes and others single lens eyes. Kirschfeld (1976) has pointed out that the resolution of a single lens eye increase proportionally to its linear dimension whilst the size of a compound eye must increase with the square of the desired resolution, thus becoming prohibitively large for the resolution achieved by vertebrate eyes. However, single lens eyes can be much smaller than compound eyes of equivalent performance (Hardie and Duelli 1978). Land (1989) concludes that the apposition eye is simply a bad design! Other suggestions include: the short focal length of compound eyes which gives rise to excellent depth of focus, even at very short distances, and also good transmission of ultraviolet light (note however, that some birds and fish also possess UV vision). Panoramic vision has also been suggested (Wehner 1981), but many vertebrates (e.g. ducks) also have wrap-round vision with a single lens eye.
70. *(p. 127)* From counts of cones and optic nerve fibres, von Fleischl, a friend of Exner's, had realised that, particularly in the peripheral retina, many cones must converge onto single optic nerve axons. He made the interesting suggestion that the convergence was interleaved, i.e. neighbouring cones would converge onto different ganglion cells. The result would be that, despite the convergence, sequential stimulation of neighbouring cones could give rise to perception of movement – although there would be uncertainty as to exactly *where* the movement had occurred. Sequential stimulation of two closely separated points in the peripheral retina does in fact give rise to the perception of a (larger) movement, even though the two points are only perceived as one when flashed simultaneously. However, since the illusion (the "fine grain movement illusion") can be elicited dichoptically it cannot depend entirely on retinal interactions (Foster et al. 1981).
71. *(p. 127)* This discussion on form and motion vision is obviously flawed. In both vertebrates and invertebrates there are similar relationships between receptor acuity and receptor spacing. Even if we accept Exner's premise with respect to overlap it is still better to have the maximum contrast transfer in single receptors. Nevertheless, this argument has pervaded the literature and constantly crops up in even quite recent text-books.
72. *(p. 128)* Plateau's experiments on flower recognition in insects were sorely lacking in appropriate controls and have since been largely discredited (for discussion see Wehner 1981).
73. *(p. 129)* Lubbock (1882) believed he had proven colour vision in bees by training them to coloured papers, however, von Frisch (1914) was the first to provide a rigorous demonstration by using a range of grey tones to control for intensity cues. One of Lubbock's major achievements was the demonstration of sensitivity to ultraviolet light in ants. In his main profession as a parliamentarian he is also accredited with the invention of the British institution of the Bank Holiday!
74. *(p. 129)* Indeed, no accommodation mechanisms have been described in compound eyes.
75. *(p. 130)*. No reports of binocular depth perception exist for stalk-eyed crustaceans. Apart from motion parallax, an alternative mechanism has recently been suggested in crabs with long eye-stalks which live in an intertidal mud-flat type habitat (Zeil et al. 1986; revs. Collett 1987; Schwind 1989). These eyes possess an acute zone aligned with the horizon. The interommatidial angles are particularly small in the vertical direction, and increase ventrally such that an object on the ground is seen by the same number of ommatidia, irrespective of distance. Zeil et al. conclude that the design allows the crabs to estimate distance by determining the point of intersection of an object with the ground – using the fixed height of the eye above the ground to complete the triangulation.

76. *(p. 130)* Most of the mechanisms of depth perception have also been uncovered in compound eyes; including motion parallax, absolute size and true stereopsis (see Rossel 1983; revs. Collett 1987; Schwind 1989)

77. *(p. 130)* The most extreme example concerns the diopsids, the stalk-eyed flies, the males of which have their eyes at the ends of long stalks generating interocular distances as large as 1 cm (Burkhardt and de la Motte 1983). Burkhardt and de la Motte (1985) argue that whilst this development may have started as a means of improving depth perception, the extreme cases are in fact a result of sexual selection.

Chapter XI

78. *(p. 132)* Exner's arguments are reasonable as far as they go. The transition between apposition and superposition eyes appears to have been made on several occasions. Plausible evolutionary scenarios are discussed at length in the light of new optical types of eyes by Nilsson (1989).

Appendix I Optical Mechanisms in Compound Eyes

Exner recognized and defined two major categories of compound eyes: those with an apposition image and those producing a superposition image. Whilst these two categories have stood the test of time, a number of distinct subtypes have been discovered in the last 20 years. Although all of these can still be assigned to either the apposition or superposition type, the optical (and in one case neural) mechanisms involved are sufficiently novel to warrant the creation of distinct categories. The following brief descriptions cover all six major optical mechanisms currently recognized, the figures are taken from the excellent recent review of Dan-Eric Nilsson (Nilsson 1989).

A. Apposition Eyes

(i) Simple (Focal) Apposition (Figs. 1a, b)

This is the simplest design, essentially the same as described by J. Müller (1826), and in fact even earlier by Hodierna (1644) and Hooke (1665). It is found very widely amongst diurnal insects and crustaceans. For historical discussion and further references see Wehner (1981) and Nilsson (1989).

Parallel rays (arrow) from a distant point source are focused by the axial corneal facet and crystalline cone (CC) onto the rhabdom (Rh). Rays entering neighbouring facets (small arrow) are absorbed by screening pigment. Fig 1b schematizes the action of the cornea as a lens (L) and the rhabdom as a waveguide (W).

(ii) Afocal Apposition (Figs. 1a, c, d)

This novel type, recently discovered in butterflies (Nilsson et al. 1984) differs only in the way light is guided to the rhabdoms (Fig 1c, d). The combination of corneal lens and crystalline cone operates as a magnifying astronomical telescope. The corneal lens focuses incident rays near the base of the crystalline cone where a powerful lens cylinder recollimates the beam into a narrow pencil of parallel rays (hence afocal). The angular magnification means that slightly off-axis light (1°) will be bent sharply (6.4°) and will not be captured by the rhabdom. A wave-optical analysis (van Hateren and Nilsson 1988) suggests that this design is slightly

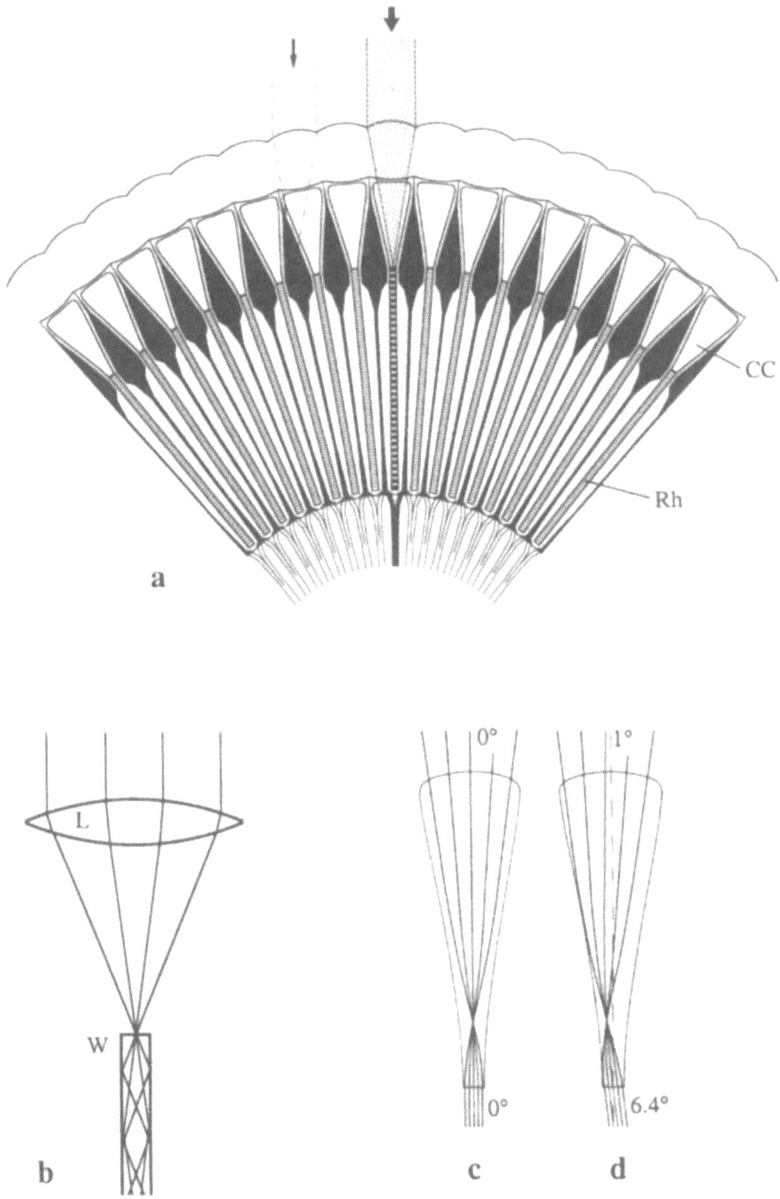

Fig. 1. *a* Apposition eye; *b* in focal apposition eyes rays are focused on the entrance of the waveguide (rhabdom); *c, d* In afocal apposition eyes a powerful lens cylinder in the proximal part of the cone recollimates the rays.

superior to conventional focal apposition eyes in terms of both absolute sensitivity and angular sensitivity.

(iii) Neural Superposition (Fig. 2)

The principle of neural superposition was in fact foreseen by Vigier (1907 a, b, c; translation in Braitenberg and Strausfeld 1973) but only recently was it rediscovered and put on a firm experimental basis (Kirschfeld 1967; Braitenberg 1967). It is found in all the higher dipterans, and in a less refined form in certain aquatic hemipterans. The secret of the system lies in the open rhabdomere arrangement, whereby seven rhabdomere tips are disposed in the focal plane of the lens in a precise geometrical array which mimics the array of facets. Parallel rays (arrows) entering neighbouring facets (only three shown) will be focused on a different rhabdomere in each ommatidium. The axons of precisely these rhabdomeres converge on the same interneurones at the first synapse where their signals are summed. In principle the same light capture could be achieved by simply having one large rhabdom under each lens, but this would throw away acuity. The beauty of this system is that light capture can be increased by up to a factor 7 without any loss of acuity. Since each ommatidium remains optically isolated, the eye is still basically of the apposition type.

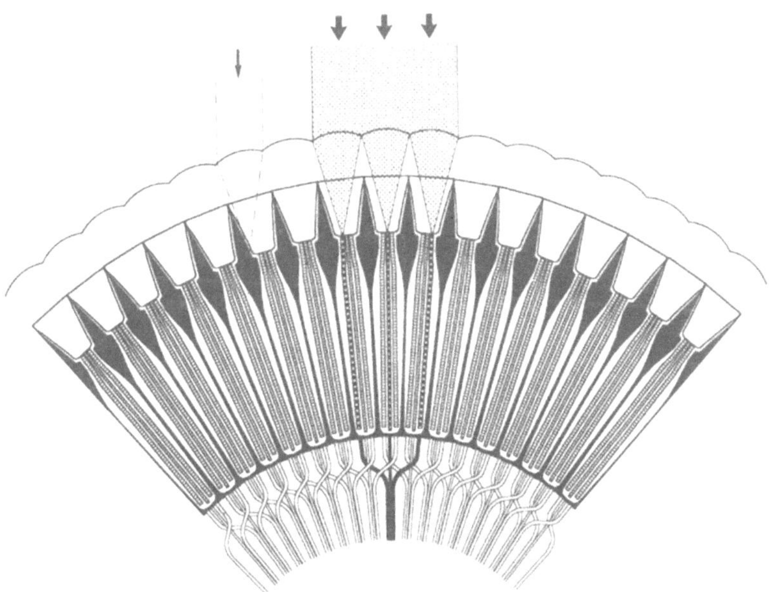

Fig. 2. Neural superposition

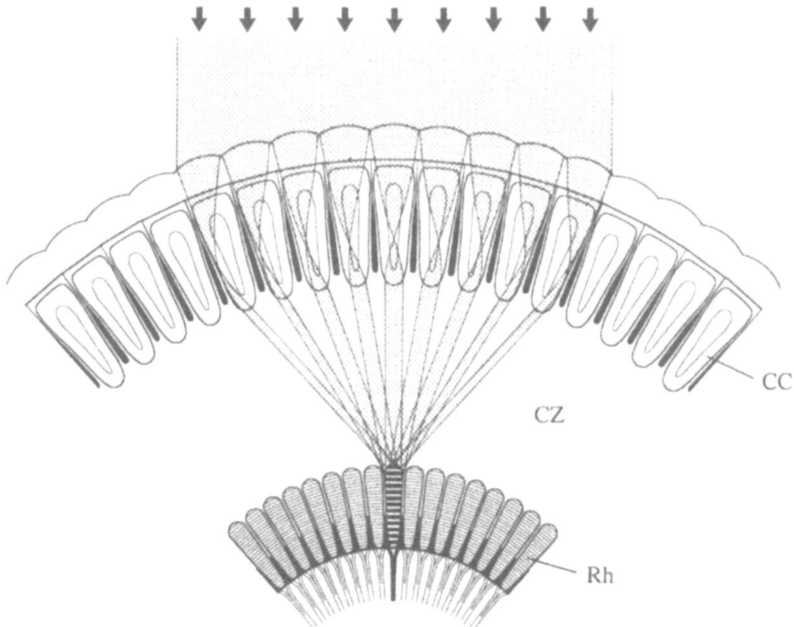

Fig. 3. Classical (refracting) superposition

B. Superposition Eyes

(i) Classical Refracting Superposition (Fig. 3)

This is the mechanism discovered and described by Exner. Although it came under serious challenge in the 1960's, the mechanism is now believed to operate amongst a wide range of nocturnal insects and crustaceans more or less exactly as Exner described it. Parallel rays are collected via a large number of facets and redirected across the clear zone (CZ) onto (ideally) a single axial rhabdomere by a combination of spherical refracting boundaries, and most importantly, the lens cylinder construction of the crystalline cones (CC). The absolute sensitivity of the system is greatly increased relative to the apposition design. Although acuity normally suffers in such eyes, Land (1984) has shown that the superposition eyes of some day-flying moths can approach the resolution of typical diurnal apposition eyes.

(ii) Reflecting Superposition (Fig. 4)

This class of eye was discovered only recently (Vogt 1975, and independently by Land 1976). In many ways it is similar to the classical superposition eye, the difference being that rays are *reflected* across the clear zone by internal reflection

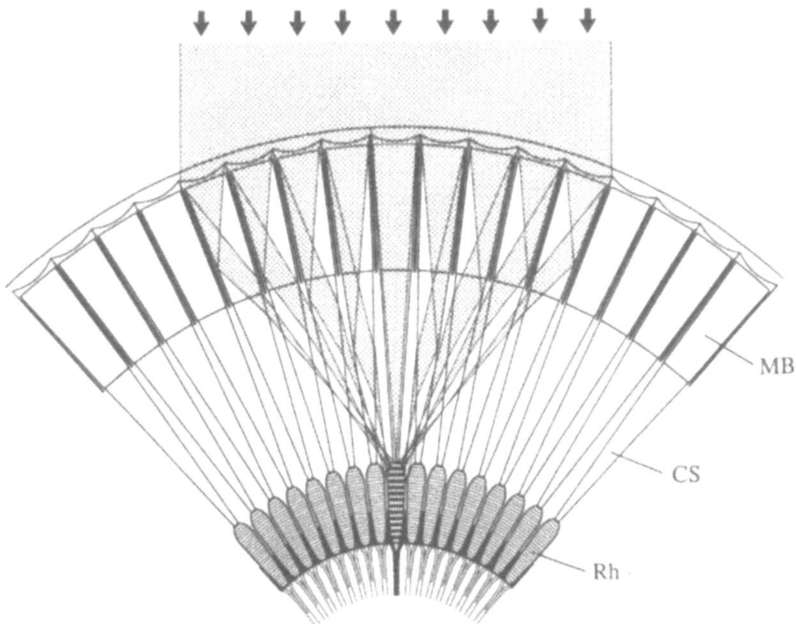

Fig. 4. Reflecting superposition

within the crystalline cones, which act as mirror boxes (MB). Such eyes are found in many decapod crustaceans, particularly amongst the macrurans. It is essential that the mirror boxes are square if the reflected rays are to reach the axial rhabdom and hence the most distinctive features of such eyes are their square facets and crystalline cones. For rays near the axis reflection also occurs within the cone stalks (CS). A precisely tailored refractive index gradient along the length of the cone stalks ensures that only the appropriate paraxial rays undergo total internal reflection, whilst more oblique rays crossing the clear zone pass through.

(iii) Parabolic Superposition (Fig. 5)

This is the most recent addition to the list of optical mechanisms (Nilsson 1988, 1989). It seems to be quite widespread amongst swimming crabs and hermit crabs and maybe even some insects. In addition many species have eyes intermediate between apposition and parabolic superposition types.

As in the reflecting superposition type eye there is no lens cylinder gradient in the crystalline cone and light is again redirected across the clear zone by reflection. However, in these eyes the corneal facets are powerful lenses and focus axial light onto light guide (LG) structures (prolongations of the retinula cells) which guide axial light to the rhabdoms. Non-axial light, however, is reflected and

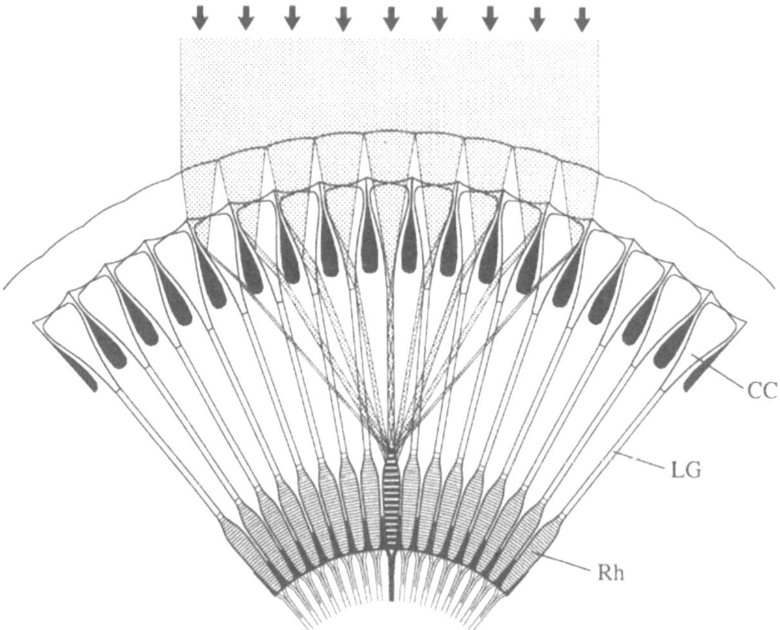

Fig. 5. Parabolic superposition

recollimated by the walls of the crystalline cones which conceal a graceful parabolic curvature and a multi-layer reflector consisting of regularly arranged endoplasmic reticulum. The parabolic curvature, which along with the presence of light guides, is a characteristic feature of these eyes, is necessary to cancel the convergence caused by the corneal lens.

Appendix II Exner's Species

For the sake of historical accuracy, the scientific names used by Exner in his original text have been faithfully reproduced. The following list includes all 93 species names used by Exner along with their presumed modern equivalents where different, and common name where appropriate. Family names are only included for Diptera, Lepidoptera, Coleoptera and Orthoptera. In addition the type of optical mechanism used by each eye is indicated (refer to App. I):

Ap	apposition
Ap(a)	afocal apposition
Ap(p)	transitional form between apposition and parabolic superposition
Ns	neural superposition
Sup	superposition (Exner's refracting type)
Sup(r)	reflecting superposition
Sup(p)	parabolic superposition
Naup	Nauplius scanning eye

Name in text	Modern name	Family	Common name	Eye type
1. Crustaceans				
Astacus fluviatilis	*A. astacus*		freshwater crayfish	*Sup(r)*
Copilia	/		copepod	*Naup*
Crangon	/		shrimp	*Sup(r)*
Carcinus maenas	/		shore crab	*Ap(p)*
Dromia	/		sponge crab	*Sup(p)*
Galathea strigosa	/		squat lobster	*Sup(r)*
Gammarus roeselli	/		amphipod	*Ap*
Homarus	/		lobster	*Sup(r)*
Innachus	*Inachus*		spider crab	*Ap(p?)*
Limulus polyphemus	/		horseshoe crab	*Ap*
Maja verrucosa	*Maia v.*		small spider crab	*Ap(p)*
Mysis	/		opossum shrimp	*Sup*
Nica edulis	*Processa e.*			*Sup(r)*
Pagurus	/		hermit crab	*Ap(p)*
Palaemon	/		prawn	*Sup(r)*
Palinurus	/		spiny lobster	*Sup(r)*
Peneus membranaceus	*Penaeus m.*		shrimp	*Sup(r)*
Phakops fekundus	/		trilobite	?
Phronima	/		amphipod	*Ap*
Pisa	/		crab	*Sup(p)*
Portunus	*Macropipus*		swimming crab	*Sup(p)*

(continued)

Name in text	Modern name	Family	Common name	Eye type
Sapphirina	/		copepod	*Naup*
Scyllarus arctus	/		bulldozer shrimp	*Sup(r)*
Sicyonia sculpta	/		shrimp	*Sup(r)*
Squilla mantis	/		mantis shrimp	*Ap*
2. Insects				
a) Odonata				
Aeschna	*Aeshna*		dragonfly	*Ap*
Agrion	/		damselfly	*Ap*
Cordulegaster	/		dragonfly	*Ap*
Libellula depressa	/		dragonfly	*Ap*
Libellula vulgata	*Sympetrum vulgatum*		dragonfly	*Ap*
b) Ephemeroptera				
Chloe binoculata	*Baetis bioculatus*		mayfly	*Ap*
c) Orthoptera				
Locusta caudata	*Tettigonia c.*	Tettigoniid	eastern great green bush cricket	*Ap*
Locusta viridissima	*Tettigonia v.*	Tettigoniid	great green bush cricket	*Ap*
Psophus stridulus	/	Acridiid	grasshopper	*Ap*
d) Neuroptera				
Ascalaphus	/		owlfly	*Sup*
e) Lepidoptera				
Acherontia atropos	/	Sphingid	deaths-head hawkmoth	*Sup*
Argynnis latonia	*A. lathonia*	Nymphalid	Queen of Spain fritillary	*Ap(a)*
Catocala nupta	/	Noctuid	red underwing	*Sup*
Cnophria quadra	*Lithosis quadra*	Arctiid	four-spotted footman	*Sup*
Coeneonympha pamphilus	/	Satyrid	small heath	*Ap(a)*
Coleas hyale	*Colias h.*	Pierid	pale clouded yellow	*Ap(a)*
Colias rhamni	*Gonepteryx r.*	Pierid	brimstone	*Ap(a)*
Epinephele	*Maniola*	Satyrid	meadow brown	*Ap(a)*
Euprepria	*Arctia*	Arctiid	tigermoth	*Sup*
Hesperia coma	*H. comma*	Hesperiid	silver-spotted skipper	*Sup*
Hipparchia medea	*Erebia aethiops*	Satyrid	scotch argus	*Ap(a)*
Lasiocampa quercifolia	*Gastropacha q.*	Lasiocampid	lappet moth	*Sup*
Leucoma salicis	/	Lymantriid	white satin moth	*Sup*
Lycaena corydon	*Lysandra cordion*	Lycaenid	chalk-hill blue	*Ap(a)*
Makroglossa stellatarum	*Macroglossum s.*	Sphingid	humming bird hawkmoth	*Sup*
Melanargia galathea	/	Satyrid	marbled white	*Ap(a)*
Ocneria rubea	/	Lymantriid	(European moth)	*Sup*
Pararge megarea	*Lasiommata megera*	Lymantriid	wall butterfly	*Sup*
Pieris brassicae	/	Pierid	large white	*Ap(a)*
Pieris crataegi	*Aporia c.*	Pierid	black-veined white	*Ap(a)*
Pieris rapae	/	Pierid	small white	*Ap(a)*
Plusia chrysetis	*Diachrysia chrysitis*	Noctuid	burnished brass	*Sup*
Plusia gamma	*Autographa g.*	Noctuid	silver Y	*Sup*
Polias	*Polia*	Noctuid		*Sup*

(continued)

Name in text	Modern name	Family	Common name	Eye type
Polyomatus phlaeas	*Lycaena p.*	Lycaenid	small copper	*Ap(a)*
Porthesia	*Euproctis*	Lymantriid		*Sup*
Sphinx convolvuli	*Agrius c.*	Sphingid	convolvulus hawkmoth	*Sup*
Spilosoma	/	Arctiid		*Sup*
Vanessa c-album	*Polygonia c-*	Nymphalid	comma	*Ap(a)*
Vanessa cardui	*Cynthia c.*	Nymphalid	painted lady	*Ap(a)*
Vanessa urticae	*Aglias u.*	Nymphalid	small tortoiseshell	*Ap(a)*
f) Hymenoptera				
Bombus terrestris	/		buff-tailed bumblebee	*Ap*
Sirex gigas	*Urocerus g.*		wood-wasp	*Ap*
Tenthredo instabilis	/		sawfly	*Ap*
Vespa crabro	/		hornet	*Ap*
g) Diptera				
Eristalis	/	Syrphid	hoverfly	*Ns*
Musca domestica	/	Calliphorid	housefly	*Ns*
Musca vomitoria	*Calliphora v.*	Calliphorid	blowfly	*Ns*
Stratiomys	/	Stratimoyid	soldier fly	*Ns*
Volucella	/	Syrphid	hoverfly	*Ns*
h) Coleoptera				
Agrypnus moertus	*A. aequalis*	Elaterid	(Indian species)	*
Cantharis fusca	/	Cantharid	soldier beetle	*
Cantharis rustica	/	Cantharid	soldier beetle	*
Cetonia affinis	/	Scarabeid	rose chafer	*Sup*
Cetonia metallica	*C. cuprea*	Scarabeid	rose chafer	*Sup*
Colymbetes fuscus	/	Dytiscid	diving beetle	*Sup*
Dorcadion aethiops	/	Lamiid		*Ap*
Dytiscus marginalis	/	Dytiscid	diving beetle	*Sup*
Elater noctilucus	/	Elaterid	click beetle	*
Hydrophilus piceus	/	Hydrophilid	water beetle	*Sup*
Lampyris splendidula	*Lamprohiza s.*	Lampyrid	firefly	*Sup*
Luciola italica	/	Lampyrid	firefly	*Sup*
Orycetes rhinoceros	*Orcytes r.*	Dynastid	rhinoceros beetle	*Sup*
Rhagonycha melanura	/	Cantharid	soldier beetle	*
Telephorus	*Cantharis*	Cantharid	soldier beetle	*
Trichodes apiarius	/	Clerid		*Ap*
Tropinota hirtella	*T. hirta*	Scarabeid	rose chafer	*Sup*

* click beetles and soldier beetles include many transitional forms between apposition and superposition.

References to Notes and Appendices

Autrum H (1981) Light and dark adaptation in invertebrates. In: Handbook of sensory physiology, vol VII/6C. Autrum H (ed) pp 1–91. Springer, Berlin Heidelberg New York

Barlow H (1952) The size of ommatidia in apposition eyes. J Exp Biol 29: 667–674

Barlow R, Chamberlain SC, Lehman HK (1989) Circadian rhythms in the invertebrate retina. In: Facets of vision. Stavenga DG, Hardie RC (eds) Springer, Berlin Heidelberg New York Tokyo

Bernard GD, Miller WH (1968) Interference filters in the corneas of Diptera. Invest Ophthalmol 7: 416–434.

Bidder F (1839) Zur Anatomie der Retina, insbesondere zur Würdigung der Stabförmigen Körper in derselben. Arch Anat Physiol Wiss Med (J Müller) pp 371–388

Braitenberg V (1967) Patterns of projection in the visual system of the fly. I. Retina–lamina projcetions. Exp Brain Res 3: 271–298

Braitenberg V, Strausfeld NJ (1973) Principles of the mosaic organization in the visual system's neuropil of *Musca domestica* L. In: Handbook of sensory physiology, vol VII/3. Autrum H (ed) pp 631–660. Springer Berlin Heidelberg New York

Boll F (1876) Zur Anatomie und Physiologie der Retina. Mon Ber Berlin Akad 41: 783–787

Brücke E (1844) Über die physiologische Bedeutung der stabförmigen Körper unter den Zwillingszapfen in den Augen der Wirbelthiere. Arch Anat Physiol 444–451

Burkhardt D, de la Motte I (1983) How stalk–eyed flies eye stalk–eyed flies: observations and measurements of the eyes of *Cyrtodiopsis whitei* (Diopsidae, Diptera). J Comp Physiol A 151: 407–422

Burkhardt D, de la Motte I (1985) Selective pressures, variability, and sexual dimorphism in stalk–eyed flies (Diopsidae). Naturwissenschaften 72: 204–206

Caveney S, McIntyre P (1981) Design of graded–index lenses in the superposition eyes of scarab beetles. Phil Trans R Soc London Ser B 294: 589–635

Clarkson ENK, Levi–Setti (1975) Trilobite eyes and the optics of Descartes and Huygens. Nature (London) 254: 663–667

Collett TS (1987) Binocular depth vision in arthropods. Trends Neurosci 10: 1–2

Danneel R, Zeutschel B (1957) Über den Feinbau der Retinula bei *Drosophila melanogaster*. Z Naturforsch 12b: 580–583

de Vries HL (1956) Physical aspects of sense organs. Prog Biophys 6: 207–264

Demoll R (1917) Die Sinnesorgane der Arthropoden, ihr Bau und ihre Funktion. Vieweg, Braunschweig

Dietrich W (1909) Die Facettenaugen der Dipteren. Z Wiss Zool 92: 465–539

Doughtie DG, Rao KR (1984) Ultrastructure of the eyes of the grass shrimp, *Palaemonetes pugio*. General morphology, and light and dark adaptation at noon. Cell Tiss Res 238: 271–288

Doujak FE (1984) Electrophysiological measurement of photoreceptor membrane dichroism and polarization sensitivity in a Grapsid crab. J Comp Physiol A 154: 597–605

Exner S (1886) Nachtrag zu der Abhandlung über "Cylinder, welche optische Bilder entwerfen". Pflüger's Arch Ges Physiol 39: 244–245

Fernandez–Moran H (1956) Fine structure of the insect retinula as revealed by electron microscopy. Nature (London) 177: 742–744

Fletcher A, Murphy T, Young A (1954) Solutions of two optical problems. Proc R Soc London Ser A 223: 216–225
Foster DH, Thorson J, McIlwain JT, Biederman–Thorson M (1981) The fine–grain movement illusion: a perceptual probe of neuronal connectivity in the human visual system. Vision Res 21: 1123–1128
Franceschini N (1975) Sampling of the visual environment by the compound eye of the fly: Fundamentals and applications. In: Photoreceptor optics. Snyder AW Menzel R (eds) pp 97–125. Springer, Berlin Heidelberg New York
Franceschini N, Kirschfeld K (1971) Les phénomènes de pseudopupille dans l'oeil composé de *Drosophila*. Kybernetik 9: 159–182
Frisch K von. (1914) Der Farbensinn und Formensinn der Biene. Zool J Physiol 37: 1–238
Goldsmith TH, Philpott DE (1957) The microstructure of the compound eyes of insects. J Biophys Biochem Cytol 3: 429–440
Gregory RL (1986) Odd perceptions. Methuen, London
Gregory RL, Ross HE, Moray N (1964) The curious eye of *Copilia*. Nature (London) 201: 1166
Grenacher H (1879) Untersuchungen über das Sehorgan der Arthropoden: ins besondere der Spinnen: Insecten und Crustaceen. Vandenhoek & Ruprecht, Göttingen
Hamdorf K, Hoglund G (1981) Light induced retinal screening pigment migration independent of visual cell activity. J Comp Physiol A 143: 305–309
Hardie RC (1985) Functional organization of the fly retina. In: Progress in sensory physiology vol 5. Ottoson D (ed) pp 1–79. Springer, Berlin Heidelberg New York Tokyo
Hardie RC, Duelli P (1978) Properties of single cells in the posterior–lateral eyes of jumping spiders. Z Naturforsch 33c: 156–158
Hateren JH van (1984) Waveguide theory applied to optically measured angular sensitivities of fly photoreceptors. J Comp Physiol A 154: 761–771
Hateren JH van (1985) The Stiles–Crawford effect in the eye of the blowfly, *Calliphora erythrocephala*. Vision Res 25: 1305–1315
Hateren JH van (1989) Photoreceptor optics, theory and practice. In: Facets of vision. Stavenga DG, Hardie RC (eds) Springer, Berlin Heidelberg New York Tokyo
Hateren JH van, Nilsson D–E (1988) Butterfly optics exceed the theoretical limits of conventional apposition eyes. Biol Cybernet (in press)
Hausen K (1973) Die Brechungsindices im Kristallkegel der Mehlmotte *Ephestia kühniella*. J Comp Physiol 82: 365–378.
Hodierna DG (1644) L'occhio della mosca. Decio Cirillo, Palermo
Höglund G (1965) Pigment migration and retinular sensitivity. In: The functional organisation of the compound eye. Bernhard CG (ed) pp. 77–101. Pergamon, Oxford
Hooke (1665) Micrographia or some physiological descriptions of minute bodies made by magnifying glasses. Martin & Allestry, London
Horridge GA (1969) The eye of the firefly *Photuris*. Proc R Soc London Ser B 171: 445–463
Horridge GA (1971) Alternatives to superposition images in clear zone compound eyes. Proc Roy Soc London Ser B 179: 97–124.
Horridge GA (1975) Arthropod receptor optics. In: Photoreceptor optics. Snyder AW, Menzel R (eds) pp 459–478. Springer, Berlin Heidelberg New York
Horridge GA (1977) Insects which turn and look. Endeavour 1: 7–17
Horridge GA, Giddings C (1971a) Movement on dark–light adaptation in beetle eyes of the neuropteran type. Proc R Soc London Ser B 179: 73–85
Horridge GA, Giddings C (1971b) The retina of *Ephestia* (Lepidoptera) Proc R Soc London Ser B 179: 87–95
Horridge GA, Marcelja, Jahnke R, McIntyre P (1983) Daily changes in the compound eye of a beetle (*Macrogyrus*). Proc R Soc London Ser B 217: 262–285
Hughes A (1977) The topography of vision in mammals of contrasting life styles. Comparative optics and retinal organisation. In: Handbook of sensory physiology, vol VII/5. Crescitelli F (ed) pp 613–756. Springer, Berlin Heidelberg New York
Kirschfeld K (1967) Die Projektion der optische Umwelt auf das Raster der Rhabdomere im Komplexauge von *Musca*. Exp Brain Res 3: 248–270

Kirschfeld K (1969) Absorption properties of photopigments in single rods cones and rhabdomeres. In: Processing of optical data by organisms. Reichardt W (ed) pp 116–136. Academic Press, New York
Kirschfeld K (1976) The resolution of lens and compound eyes. In: Neural Principles in Vision. Zettler F, Weiler R (eds) pp. 354–370. Springer, Berlin Heidelberg New York
Kirschfeld K, Franceschini N (1969) Ein Mechanismus zur Steuerung des Lichtflusses in den Rhabdomeren des Komplexauges von *Musca*. Kybernetik 6: 13–22
Kirschfeld K,Vogt K (1980) Calcium ions and pigment migration in fly photoreceptors. Naturwissenschaften 67: 516–517
Kühne W (1878) On the photochemistry of the retina and on visual purple. McMillan, London
Kuiper JW (1962) The optics of the compound eye. Symp Soc Exp Biol 16: 58–71
Kuiper JW (1966) On the image formation in a single ommatidium of the compound eye in Diptera. In: The functional organization of the compound eye. Bernhard CG (ed) Wenner-Gren Center Intern Symp
Kunze P (1979) Apposition and superposition eyes. In: Handbook of sensory physiology, vol VII/6A. Autrum H (ed) pp 441–502. Springer, Berlin Heidelberg New York
Land MF (1976) Superposition images are formed by reflection in the eyes of some oceanic decapod crustacea. Nature (London) 263: 764–765
Land MF (1980) Compound eyes: old and new optical mechanisms. Nature (London) 287: 681–685
Land MF (1981a) Optics and vision in invertebrates. In: Handbook of sensory physiology, vol VII/6B. Autrum H (ed) pp 471–592. Springer, Berlin Heidelberg New York
Land MF (1981b) Optics of the eyes of *Phronima* and other deep-sea amphipods. J Comp Physiol A 145: 209–226
Land MF (1984) The resolving power of diurnal superposition eyes measured with an ophthalmoscope. J Comp Physiol A 154: 515–533
Land MF (1986) Screening pigment migration in a sphingid moth is triggered by light near the cornea. J Comp Physiol A 160: 355–357
Land MF (1989) Variations in the structure and design of compound eyes. In: Facets of vision. Hardie RC Stavenga DG (eds). Springer, Berlin Heidelberg New York Tokyo
Langer H (1965) Nachweis dichroitischer Absorption des Sehfarbstoffes in den Rhabdomeren des Insektenauges. Z Vergl Physiol 51: 258–263
Lubbock J (1882) Ants, bees, and wasps. (New edn by Myers JG, Dutton, New York 1929)
Mallock A (1894) Insect sight and the defining power of composite eyes. Proc R Soc London Ser B 55: 85–90
Marshall NJ (1988) A unique colour and polarization vision system in Mantis shrimps. Nature (London) 333: 557–560
Miller WH (1957) Morphology of the compound eye of *Limulus*. Biophys Biochem Cytol 3: 421–428
Miller WH (1979) Ocular optical filtering. In: Handbook of sensory physiology, vol VII/6A. Autrum H (ed) pp 69–143. Springer, Berlin Heidelberg New York
Miller WH, Bernard GD (1968) Butterfly glow. J Ultrastruct Res 24: 286–294
Müller H (1854) Über die entoptische Wahrnemhung der Netzhautgefässe, insbesondere als Beweismittel für die Lichtperception durch die nach hinten gelegenen Netzhautelementen. Verh Dtsch Phys Med Ges Würzburg 5: 411
Müller J (1826) Zur vergleichenden Physiologie des Gesichtsinnes des Menschen und der Thiere nebst einem Versuch über die Bewegungen der Augen und über den menschlichen Blick. Cnobloch, Leipzig
Nilsson D-E (1989) Optics and evolution of the compound eye. In: Facets of vision. Hardie RC Stavenga DG (eds). Springer, Berlin Heidelberg New York Tokyo
Nilsson D-E (1988) A new type of imaging optics in compound eyes. Nature (London) 332: 76–78
Nilsson D-E, Land MF, Howard J (1987) Afocal apposition optics in butterfly eyes. Nature (London) 312: 561–563
Nowikoff M (1931) Untersuchungen über die Komplexaugen von Lepidopteren nebst einigen Bemerkungen über die Rhabdome der Arthropoden im allgemeinen. Z Wiss Zool 138: 1–67

Pictet FJ (1843) Histoire naturelle générale et particulière des Insects Neuroptères. Famille des Ephemerinés. Kessmann & Cherbuliez, Geneva
Rao KR (1985) Pigmentary effectors. In: The biology of Crustacea vol 9. Bliss DE (ed) pp 395-462. Academic Press, Orlando New York
Ribi WA (1978) Ultrastructure and migration of screening pigments in the retina of *Pieris rapae* L. (Lepidoptera, Pieridae). Cell Tissue Res 191: 57-73
Rossel S (1983) Binocular stereopsis in an insect. Nature (London) 302: 821-822
Schultze M (1868) Untersuchungen über die zusammengesetzten Augen der Krebse und Insecten. Cohen, Bonn
Schwind R (1989) Size and distance perception in compound eyes. In: Facets of vision. Stavenga DG, Hardie RC (eds) Springer Berlin Heidelberg New York Tokyo
Seitz G (1968) Der Strahlengang im Appositionsauge von *Calliphora erythrocephala* Meig. Z Vergl Physiol 59: 205-231
Seitz G (1969) Untersuchungen am dioptrischen Apparat des Leuchtkäferauges. Z Vergl Physiol 62: 61-74
Snyder AW (1977) Acuity of compound eyes: physical limitations and design. J Comp Physiol A 116: 161-182
Snyder AW (1979) Physics of vision in compound eyes In: Handbook of sensory physiology, vol VII/6A. Autrum H (ed) pp 225-313. Springer, Berlin Heidelberg New York
Snyder AW, Menzel R (eds) (1975) Photoreceptor optics. Springer, Berlin Heidelberg New York
Stavenga DG (1975) Optical qualities of the fly eye – an approach from the side of geometrical, physical and waveguide optics. In: Photoreceptor optics. Snyder AW Menzel R (eds) pp 126-144. Springer, Berlin Heidelberg New York
Stavenga DG (1979) Pseudopupils of compound eyes. In: Handbook of sensory physiology, vol VII/6A. Autrum H (ed) pp 357-439. Springer, Berlin Heidelberg New York
Stavenga DG (1989) Pigments in compound eyes. In: Facets of vision. Stavenga DG, Hardie RC (eds). Springer, Berlin Heidelberg New York Tokyo
Stockton WL, Cowen R (1976) Stereoscopic vision in one eye: palaeophysiology of the schizochroal eye of trilobites. Palaeobiology 2: 304-315
Trujillo-Cenoz O (1972) The structural organization of the compound eye in insects. In: Handbook of sensory physiology, vol VII/2. Fuortes MGF (ed) pp 5-62. Springer, Berlin Heidelberg New York
Tsuda M (1987) Photoreception and phototransduction in invertebrate photoreceptors. Photochem Photobiol 45: 915-931
Vigier P (1907a) Mécanisme de la synthèse des impressions recueillies par les yeux composés des Diptres. CR Acad Sci Paris 63: 122-124
Vigier P (1907b) Sur les terminations photoreceptrices dans les yeux composés des Muscidés. CR Acad Sci Paris 63: 532-536
Vigier P (1907c) Sur la reception de l'excitant lumineux dans les yeux composés des Insectes, en particulier chez les Muscidés. CR Acad Sci Paris 63: 633-636
Vogt K (1974) Optische Untersuchungen an der Cornea der Mehlmotte *Ephestia kühniella*. J Comp Physiol 88: 201-216
Vogt K (1975) Zur Optik des Flusskrebsauges. Z Naturforsch 30: 691
Vogt K (1977) Ray path and reflection mechanisms in crayfish eyes. Z Naturforsch 32c: 466-468
Vogt K (1980) Spiegeloptik des Flusskrebsauges. The optical system of the crayfish eye. J Comp Physiol A 135: 1-19
Walcott B (1975) Anatomical changes during light-adaptation in insect compound eyes. In: The compound eye and vision of insects. Horridge GA (ed) pp. 20-33. Clarendon, Oxford
Walls (1942) The vertebrate eye and its adaptive radiation. Hafner, New York
Waterman TH (1981) Polarization sensitivity. In: Handbook of sensory physiology Vol VII/6B. Autrum H (ed) pp. 283-469. Springer, Berlin Heidelberg New York
Wehner R (1981) Spatial vision in insects. In: Handbook of sensory physiology, vol VII/6C. Autrum H (ed) pp 287-616. Springer, Berlin Heidelberg New York
Wolken JJ, Florida RG (1969) The eye structure and optical system of the crustacean copepod *Copilia*. J Cell Biol 40: 279-285
Wolken JJ, Capenos J, Turano A (1957) Photoreceptor structures III *Drosophila melanogaster*. J Biophys & Biochem Cytol 3: 441-448

Yagi N, Koyama N (1963) The compound eye of Lepidoptera. Shinkyo, Tokyo
Zeil J, Nalbach G, Nalbach H-O (1986) Eyes, eye-stalks and the visual world of semi-terrestrial crabs. J Comp Physiol A 159: 801-811
Zenker W (1867) Versuch einer Theorie der Farben-Perception. Arch Mik Anat 3: 248-261

In addition the reader may find interesting information and further references in the following review volumes, published since the revival in interest in the compound eye in the 1960's (listed chronologically).

1962. Biology of receptor organs. Symp Soc Exp Biol XVI

1966. The functional organisation of the compound eye. Bernhard CG (ed). Wenner-Gren Symp Ser, vol 7

1969. Processing of optical data by organisms. Reichardt W (ed). Academic Press, New York

1972. Information processing in the visual system of arthropods. Wehner R (ed). Springer, Berlin Heidelberg New York

1973. Biochemistry and physiology of visual pigments. Langer H (ed). Springer, Berlin Heidelberg New York

1975. Photoreceptor optics. Snyder AW, Menzel R (eds). Springer, Berlin Heidelberg New York

1975. The compound eye and vision of insects. Horridge GA (eds). Clarendon, Oxford

1976. Neural principles in vision. Zettler F, Weiler R (eds). Springer, Berlin Heidelberg New York

1979. Handbook of sensory physiology, vol VII/6A. Autrum H (ed). Springer, Berlin Heidelberg New York

1979. Handbook of sensory physiology, vol VII/6B. Autrum H (ed). Springer, Berlin Heidelberg New York

1981. Handbook of sensory physiology, vol VII/6C. Autrum H (ed). Springer, Berlin Heidelberg New York

1983. The biology of photoreception. Cosens DJ, Vince-Price D (eds). Symp Soc Exp Biol XXXVI

1984. Photoreception and vision in invertebrates. Ali M (ed). Plenum, New York

1986. Comprehensive insect physiology, biochemistry and pharmacology. Kerkut GA, Gilbert LI (eds) Vol VI: Nervous system: sensory. Pergamon, Oxford

1989. Facets of vision. Stavenga DG, Hardie RC (eds). Springer, Berlin Heidelberg New York Tokyo

Index

Index. (Translator's note: this is basically Exner's original index, but with a few additions. In particular an author index has also been included).

FSC
www.fsc.org
MIX
Papier aus verantwortungsvollen Quellen
Paper from responsible sources
FSC® C105338

If you have any concerns about our products,
you can contact us on
ProductSafety@springernature.com

In case Publisher is established outside the EU,
the EU authorized representative is:
Springer Nature Customer Service Center GmbH
Europaplatz 3, 69115 Heidelberg, Germany

Printed by Libri Plureos GmbH
in Hamburg, Germany